PRAISE FOR VOODO

"*Voodoo Vintners* is a rare thing: a book and great storytelling in such an intoxicating fashion it's hard to put down. It's a superb read, perfect for anyone interested in biodynamics, winemaking, or simply stories of people willing to color outside the lines."
—Ann Zimmerman, Serious Eats (seriouseats.com)

"This is a great book … Cole brings a light, gently witty twist to her writing and acts as an astute observer of the growers in Oregon who are turning to this alternative form of growing wine grapes."
—Jamie Goode, wineanorak.com

"As a field report from Oregon's thriving sustainable wine country, this book delivers. What's more is that Cole's book is not just aimed at wine geeks (though it will certainly appeal to them). *Voodoo Vintners* is an enjoyable—and necessary—read for anyone who might want to take wine in a more sustainable direction, or for those readers who would like to raise a glass to the people who are already working to make it happen."
—Ryan Clark, Civil Eats (civileats.com)

"Cole's book is a sensual, smart study of the Oregon wine world and the future of agriculture. Savor it slowly."
—Kerry Newberry, *Oregon Wine Press*

"The strength of this book is the lovely writing as well as the lack of starry eyes … [Cole] gives us both sides. The takeaway is that there's more than one way to till (or not) the land … Even though the narrative is Oregon farmer based, these are universal stories, and an enjoyable read."
—Alice Feiring, The Feiring Line (alicefeiring.com)

"Like the Portlander who rides his bike every day, rain or shine, the biodynamic winegrower is a stubborn sort who courts disaster in hopes of creating a miracle. If you think biodynamic wine sounds like the latest snake oil, you'll find plenty in *Voodoo Vintners* to confirm your impression. If you want a knowledgeable, engaging, boots-in the-field introduction, you couldn't find a better guide than Cole."
—Angie Jabine, *The Oregonian*

Voodoo Vintners

OREGON'S ASTONISHING BIODYNAMIC WINEGROWERS

Katherine Cole

Oregon State University Press • Corvallis

The paper in this book meets the guidelines for permanence and durability of the Committee on Production Guidelines for Book Longevity of the Council on Library Resources and the minimum requirements of the American National Standard for Permanence of Paper for Printed Library Materials Z39.48-1984.

Library of Congress Cataloging-in-Publication Data
Cole, Katherine.
 Voodoo vintners : Oregon's astonishing biodynamic winegrowers / Katherine Cole.
    p. cm.
 Includes bibliographical references and index.
 ISBN 978-0-87071-605-8 (alk. paper)
 1. Viticulture--Oregon. 2. Organic farming--Oregon. 3. Wine and wine making--Oregon. I. Title.
 SB387.76.O7C66 2011
 634.8'88409795--dc22

                              2010053121

© 2011 Katherine Claiborne Cole
All rights reserved.
First published in 2011 by Oregon State University Press
Second printing 2012
Printed in the United States of America

Oregon State University Press
121 The Valley Library
Corvallis OR 97331-4501
541-737-3166 • fax 541-737-3170
   http://oregonstate.edu/dept/press

Table of Contents

Acknowledgements

For the past decade, the winemakers and vinegrowers of Oregon have generously welcomed me into their vineyards and cellars. I'm indebted to all of them for my vinous education. I'm especially grateful to those who allowed me into their lives to gather information for this book.

I'm still wondering why the editors at *The Oregonian*, one of the last great daily newspapers, took a chance on this unknown fledgling wine writer back in 2002. I'm truly proud to be associated with this publication, and to be working with the teams at FOODday and at *MIX* magazine.

At Oregon State University Press, Mary Braun, Jo Alexander, and Micki Reaman have been instrumental in turning my loony idea into a manuscript, page proofs, and finally, a book.

The indispensible Raechel Sims, my fact checker, marketing mistress, and all-around Girl Friday, is quite possibly the most competent person on the planet.

More gratitude to Katie Mitchell and Kaitlin Cloninger, for their assistance, and to my friend Christina Henry de Tessan for her encouragement and guidance.

My parents, Douglass and Katherine Raff, have been unquestioning in their love and support and indexing (thank you, Mom!). And my husband, Peter Cole, and our children, Elizabeth and Caroline, are my *raison d'être*.

Preface

Wine writers play two roles: that of the reporter, and that of the critic. We are expected to publish reliable information as well as knowledgeable opinions.

So let me begin by informing you what this book is not: it is not a review of wines or a compendium of useful information, such as you might find in a wine guide. Instead, it is an examination of an inscrutable topic.

I first became aware of biodynamic viticulture sometime around the year 2000, when I moved to Oregon and began tasting local wines. I remember being struck at that time by a pinot gris that was quite unlike its peers: crisp and clean, it reminded me of the pure water you might drink from a mountain spring. I would later discover that it had been made from biodynamic grapes.

A couple of years later, I met the charismatic Jimi Brooks, a figure who appears repeatedly in the following pages, and whose riesling, at that time, had that same mountain-spring purity that I had noticed earlier in the pinot gris.

Brooks was one of those wickedly funny, effortlessly likeable people who could convince just about anyone to try just about anything. As vineyard manager and winemaker for Maysara Winery and Momtazi Vineyard as well as for his own eponymous label, Brooks, he pursued biodynamic viticulture with his typical enthusiasm. Jimi convinced me and many others to take a closer look at this unusual style of agriculture.

Touring Moe Momtazi's property with Brooks, I was struck by the tumbledown appearance of the place. It looked wild and alive—so unlike the neighboring estates, with their neat vine rows of brown-and-green corduroy. As I wrote at the time, "The access road was hemmed in by swampy ditches and weed-laden mounds of percolating manure; farther up the steep, rutted alleys of Maysara's Momtazi Vineyard, sheep, chickens, cows, and horses ambled through untamed fields. Patches of brambles and poison oak harbored coveys of quail. And rambling rows of vines were accented by corridors of crimson clover and purple vetch."

I was shocked to find Brooks carefully tending stands of nettles and horsetail—in my estimation, noxious weeds. I thrilled to see him stirring these weeds into teas, using a witchy-looking twig broom, with a mischievous grin on his face.

But this was Oregon, where it's typical for a local to complain that she's having a bad day solely due to the position of Saturn in the sky. In the Oregon wine community, Brooks was just one of many off-the-wall characters making wines in an unconventional way.

Then I began reading more about biodynamic viticulture. I learned that some of France's most respected vintners were pursuing the practice, and that the goddess-like Lalou Bize-Leroy, of Domaine Leroy in Burgundy, had spoken on the subject at the International Pinot Noir Celebration in McMinnville, Oregon, in 2001. Could biodynamic viticulture be a serious, worldwide movement?

Burgundy's best vignerons were doing it. So, increasingly, were Oregon's best. I discovered that the headquarters of the American biodynamic movement, Demeter USA and the Biodynamic Farming and Gardening Association, are both based in Oregon.

(An aside here: Demeter USA has trademarked the words "Demeter" and "biodynamic" so that they don't become diluted in the manner of fuzzy terms such as "green" and "natural." These registered certification marks protect consumers, biodynamic producers, and it goes without saying, Demeter USA. If estates without Demeter certification market or label their wines as "bio-dynamic," they risk legal action for trademark infringement. I have attempted to make clear in the following pages which properties are Demeter certified and which are not. However, executive director Jim Fullmer has been kind enough to grant me fair use of the terms "biodynamic" and "biodynamics" to generally describe the farming techniques associated with this practice.)

In the ensuing years, I found myself repeatedly defining and describing biodynamic viticulture for the benefit of fellow wine lovers. Their questions and interest sent me searching for books on the topic. To my dismay, I found very few. There were gardening manuals, a valuable but encyclopedic tome by the British wine writer Monty Waldin, and the original transcripts of Rudolf Steiner's 1924 lectures on the subject. There were, also, ruminations by the quirky Loire Valley vigneron Nicolas Joly, who has been an effective spokesperson for the movement, but whose baroque verbal stylings do not exactly lend themselves to easy comprehension.

Preface

Looking over this array of dense texts, I was flummoxed. Why wasn't there a simple, readable, enjoyable book on biodynamic viticulture for the everyday wine lover to flip through and enjoy? And why shouldn't this book focus on the trend from the perspective of Oregon wine country, where the biodynamic practitioners were a colorful bunch with plenty to say? A story was forming in my mind, a story much larger than one that could be crammed into the occasional newspaper column.

It was at this time that I was approached by Mary Braun of Oregon State University Press and invited to submit a book proposal. I sent her an outline of the story of biodynamic wine in Oregon. The Press kindly accepted my proposal, and I set to work.

Over the course of the following year—during which I continued with my regular commitments for *The Oregonian* and *MIX* magazine, as well as my duties as the mother of two small children and the wife of a very busy, if very supportive, husband—I got into the habit of jumping into my car and cruising out to wine country whenever I could find an extra half-day, and ducking into my office to type whenever I could find a spare moment. By the year's end, I had this: a book about biodynamic agriculture as seen through the lens of Oregon viticulture.

Thanks to its cow horns, moon phases, and cachet, biodynamic winegrowing makes for a compelling story. But I hope this book also expresses my admiration of all the Oregonians who toil in the vineyard and tinker in the cellar, no matter what style of winegrowing they are practicing.

This book is about anyone insane enough to be buffeted by the Willamette Valley's famous rains eight months out of every twelve. It is about the organic, the sustainable, and the conventional vignerons. All are foolhardy enough to make pinot noir in Oregon; it is just one small step from this level of risk to that even more foolhardy form of farming, biodynamics.

I have an emotional response to the very best wines. Like the very best books and films, they make me weep. My regular readers know by now what sorts of wines make me cry: They're usually lower in alcohol, higher in acidity, more mineral, less ripe. They're tense and electric. And, of course, they reek of *terroir*.

I don't know if biodynamic agriculture is the key to unlocking *terroir*; I suspect a number of factors count, starting with the suitability of the site. Still, some readers might wonder what I think of biodynamic wines: Do I prefer them or dislike them? Do they make me reach for the Kleenex box? My answer is this: biodynamically farmed grapes make fascinating wines. They also make banal wines. The same is true of conventionally farmed grapes, organically farmed grapes, and everything in between.

The pragmatist in me is suspicious of the biodynamic movement. While many of its farming practices and ecological premises appear sound, they come packaged with a lot of extraneous spiritual baggage that I can't help but view cynically.

However, I must admit: as someone who drives a stick shift when she's not getting around on foot or by bike, I feel camaraderie with anyone who prefers to take the more arduous path to arrive at his or her destination. It may not be the most efficient way to get there, but it is, in my experience, always the most pleasurable.

Introduction

You do something to me,
something that simply mystifies me.
Tell me, why should it be
you have the pow'r to hypnotize me?
Let me live 'neath your spell
Do do that voodoo
that you do so well.

—Cole Porter, *You Do Something to Me*

They say you can't judge a book by its cover, nor a wine by its label. But I contend that you can tell a great deal about a winegrowing operation simply by sizing up how the winemaker is dressed: he who wears a blazer and button-down to work, for example, is probably engaged in different activities than she who dons a golf shirt and khakis.

The typical Oregon vintner's wardrobe consists of an array of Levi's, plaid flannels, Carhartts, and stained sweatshirts. All of which—although they might leave something sartorial to be desired—bode well for the wine.*

And so if, upon arriving at Brick House Vineyards in Newberg, Oregon, a visitor is greeted by a tall, broad gentleman with a salt-and-pepper beard, dressed in a faded Carhartt T-shirt and worn canvas work pants held up by an old leather belt and a tooled Western buckle, these are good omens. If this gentleman's hair should look wind-blown and wild, if that hair should be flecked with bits of straw, and if a set of neon-orange ear plugs should be dangling, sideways, around the vintner's neck, then these, too, are very good signs.

The Brick House winery is a drafty barn built in 1931, with weathered Tonnellerie Cadus wine barrels stacked against its walls. A makeshift lab (a sink, a few cabinets, rows of test tubes) is in one corner; in another, hanging tapestries conceal a desk and filing

---

* One obsessive winemaker of my acquaintance, Jay McDonald, owns seven pairs of Carhartts, in three different waist sizes. The smaller sizes cover summer and fall, when hard work happens in the vineyard and cellar; the largest are for winter and spring, with their endless wine dinners and sales trips.

cabinet. The tasting area is cozy and bunker-like, with low ceilings, an elevated, wide-planked-pine floor, and a long wood table flanked by a green enamel wood stove, antique easy chairs, and a well-worn leather couch cradling an array of snoring canines. It is a bright, sun-soaked day outside, but it is peacefully dark and churchlike inside, where barn swallows sing from the attic rafters to the slow metronome of an orchestra of ticking clocks and a Vivaldi violin concerto drifts in from concealed speakers. At the sliding-glass doors, a fat telescope points toward the moon. In short, this barn isn't merely a winery. It's the man-cave of Doug Tunnell, winegrower and proprietor of Brick House Vineyards. Tall and stalwart, he projects an Adam Bede-like nobility and speaks in a baritone with a cello-like resonance. Born and educated in Oregon, Tunnell worked as a CBS foreign correspondent, mostly based in Beirut, before handing in his press pass and purchasing a forty-acre hillside estate in the Willamette Valley.

Tunnell and his wife, Melissa Mills, live in the eponymous brick house across the drive from the barn, just past a grape arbor and a stand of trees; both structures are built into the hillside and surrounded by vines. Two decades ago, when he purchased the old hazelnut and walnut orchard and converted it to a pinot noir, chardonnay, and gamay noir vineyard, Tunnell realized that he would have to breathe, soak up, and coexist with every substance sprayed on his precious plants. He decided to transition the property to organic agriculture and began to compile a chemical history of the land, interviewing some of the sprayers who had applied pre-emergent herbicides, fungicides, and insecticides such as Paraquat and DDT. "The list of substances read like a chapter out of *Silent Spring*," he recalls.

Because he lives among his vines, Tunnell's relationship with his property is sensual. His daily existence is one of seeing brown canes and trifoil grape leaves, stepping on soft earth, smelling the green vegetation around him, tasting the ripening fruit. Which, he believes, is why he noticed, sometime around the year 2000, something slipping away. "From the day I owned the farm, we used only organic methods of maintenance. But I started to feel like I was losing some of the character of the grapes, especially the chardonnay," he recalls.

Introduction

His plants seemed to droop a bit. His soil felt brittle as he sifted it through his fingers. "The fact that we are growing on hillsides makes it even more important that we work with the soil," Tunnell reflects. "It is just so fragile. It dries out so thoroughly. It just needs nourishment. It needs microbial activity. It needs organic matter. We can't just keep mining fruit out of this place and not putting anything back in. It can't be a one-way street. Nothing in nature is."

The path to soil sickness is gradual and subtle in a fertile place like western Oregon. It might begin before a vineyard is ever planted. Earth-moving equipment uproots trees, bushes, and boulders, then turns over the soil to make it smooth. As soon as the vines are planted, weeds start to sprout up and threaten to crowd them out.

The vinetender applies herbicide, but this kills off any benign cover crops that might try to resurrect themselves, thus depleting the soil of nutrient-gathering capabilities, and clearing out the microbe-feeding buffet of decomposing vegetation.

The downed trees and shrubs and lack of a cover crop mean that roosting spots for birds and beneficial insects are gone. The result is a pest problem both large (gophers, which hawks and owls would have hunted) and small (bugs, which smaller birds and larger insects would have gotten). This calls for pesticides. Which curtail the aerating and phosphorus-releasing capabilities of the earthworm. And so pathogenic fungi, which thrive in anaerobic conditions, move in.

The soil grows brittle, then rock-hard and lifeless for lack of air. The farmer tills or discs to soften and aerate the earth, dispersing dust and breaking down whatever organic matter was left. Which leaves the soil bereft of humus, the matter that stores moisture and nutrients.

The grapevines grow droopy and begin to contract diseases and attract pests. So the vinetender applies fertilizer, which is like a steroid shot straight to the vein of the plant, pumping it up for now but setting it up for a future heart attack or stroke.

As an organic farmer, Doug Tunnell's situation wasn't as extreme as the admittedly dramatic one outlined above. But he still wasn't satisfied with the health of his vines or the fertility of his soil. So in 2000, Tunnell began to study a deeper form of agriculture called biodynamics. This style of cultivation goes beyond the organic imperative of "do no harm"; it aims to actually *improve* farm health. "The organic programs as we followed them placed some emphasis on feeding the soil, but biodynamics makes that job absolutely integral to the methodology and our soils and microbial populations have benefited greatly as a result," says Tunnell.

Tunnell describes biodynamics as "a holistic approach": "If you have healthy soil, you will have healthy plants. And if you have healthy plants, you will have better fruit. And if you have better fruit, you will have better wine. And if you have better wine, you will have better customers and happier people." By 2005, Brick House Vineyards was a Demeter-certified biodynamic property.

What looks different about a biodynamic vineyard? The rows might be a bit wilder, with red clover blossoms peeking up between the vines; or there might be bird boxes mounted above each block. But in Oregon, these are fairly common sights.

Perhaps, then, it's the rambling English garden near the winery at Brick House Vineyards. Look closely between soft pink rose blossoms and you'll see homeopathic curatives, such as bright-white chamomile, yellow and coral clumps of valerian, and gold, white, and paprika yarrow, as well as neat rows of stinging nettle—a surprising sight, since most gardeners consider nettles to be a bothersome nuisance.

Nearby is a brick-lined pit that looks like a shallow well; it will be used to make "cow-pat" or "barrel" compost. And behind the corrugated-aluminum machine shed is a row of massive compost piles. These mounds used to be nothing more than piles of shit (cow dung, actually) mixed with weeds, straw, and grapevine debris. But now, through natural alchemy, they are soft coffee-colored hills of humus, flecked with seeds, hulls, and pebbles, smelling like rich soil. Tunnell plunges both hands in and holds out a fistful of the stuff, pointing out ten red tiny wriggling worms.

Biodynamic practitioners build and nurture compost piles. They grow cover crops to fix nitrogen in the soil. They install birdhouses to attract songbirds, hawks, and owls. They send sheep and chickens down the vine rows to feast on weeds and work the earth. In these ways, they are not too different from proactive organic farmers.

Introduction

But there are two aspects of biodynamic agriculture that set it apart. Practitioners time their cultivating to the movement of the moon and the stars in the sky. And they make and apply the biodynamic preparations. The preparations, or preps, as they're often called, are like homeopathic treatments for plants. They were first listed by the Austrian scholar Rudolf Steiner in a 1924 series of lectures to farm owners—today collected in a single volume entitled *Spiritual Foundations for the Renewal of Agriculture*—that are the basis of biodynamic farming. They also offer a taste of Steiner's New-Agey credo, with their frequent references to cosmic influences and life forces.

Because the preps are what have captured the public's imagination about BD (a commonly used nickname for biodynamic agriculture) and because they are a required part of any certified-biodynamic farming regime, they will be referred to throughout this book. So let's take a moment to familiarize ourselves with them.

Preparation 500* is a cow horn packed with the manure of lactating bovines—no bullshit—and buried two and a half to five feet underground for the winter, "the season when the Earth is most inwardly alive," according to Steiner. It's dug up in the spring, by which point the manure looks like finely pulverized coffee grounds and smells and feels like soft, rich earth. In minute portions, it is added to half a bucket of water at a time. This water is stirred vigorously for an hour, in a ritualistic manner that will be described further in Chapter Three. It's sprayed on the soil in late spring and late autumn to encourage root growth.

To make Preparation 501, the farmer packs a cow horn "with quartz that has been ground to a powder and mixed with water to the consistency of a very thin dough." He buries this for the summer, digs it up in late autumn, and saves it for the following spring, when he stirs a tiny quantity (Steiner is aggravatingly vague here: "you can take a portion the size of a pea, or maybe no bigger than a pinhead")

---

* According to biodynamic author and expert Monty Waldin, the numerals are no more than product codes; the first 499 numbers were already taken by anthroposophic/homeopathic medicines produced by Weleda, Steiner's pharmaceutical company. However, Waldin points out that they were also useful as a code jargon during the Third Reich, when the practice of biodynamics was banned in Germany.

into a whole bucket of water and sprays it on foliage to promote photosynthesis and ripening.

To prepare 502, the practitioner gathers a bunch of yarrow flowers in the summer, stuffs them into the bladder of a deer, and hangs this up to dry in the sun until the autumn, when she buries the whole thing. She digs it up the next spring, then adds its contents to her compost pile. The resulting compost is supposed to "enliven the soil," encouraging the absorption of nutrients.

Chamomile flowers packed into a cow intestine make Preparation 503 sausages. These are buried for the winter, exhumed in the spring, and added to the compost pile. Chamomile is apparently just as soothing for the soil's digestion of nutrients as it is for ours, stabilizing nitrogen and stimulating plant growth.

Compressed wilted stinging nettles, buried for a whole year, then unearthed and added to the compost pile, produce 504. This should aid in the decomposition process, filtering out the bad stuff and retaining the useful stuff.

Preparation 505 is chopped-up oak bark, packed into the skull of a farm animal, and stashed somewhere very wet, such as under a gutter, under a snowpack, or even in a rain barrel, in late autumn. It is retrieved in the spring, its contents added to the compost pile as a calcium source that will—it is claimed—raise the pH of the soil and prevent or arrest disease.

Wilted dandelion heads stuffed into cow mesenteries (connective stomach tissue) and buried in late autumn make 506. Unearthed in the spring, they're added to the compost pile so as to stimulate the relationship between silica and potassium.

Valerian blossoms gathered during the summer and pressed result in a juice that is diluted and sprayed over the compost pile. This is 507, which is supposed to heat the compost and bring phosphorus to the soil.

A giant cauldron of tea steeped from silica-rich horsetail is 508. This is sprayed—along with Bordeaux mixture (the commonly used fungicide of copper sulfate and hydrated lime) and minute amounts of sulfur—on the soil in the spring to prevent and control fungal disease. Extra-credit points for those who make additional teas out of dried chamomile blossoms to combat heat stress or nettle leaves as an insect repellant.

Introduction

When discussing biodynamic agriculture, we struggle with a way to describe it in just a few words. It's über-organic. It's witchcraft farming. It's voodoo in the vineyard. It's all of these things, and none of these things.

The wine writer Matt Kramer likens biodynamic to kosher, dubbing Demeter, the biodynamic education and certification organization, the rabbinate that legislates this agricultural orthodoxy. "For this observer, biodynamic processes are a form of discipline, some of which may well actually work, while other practices may be more emotionally and psychologically sustaining to the practitioner than practical to the plant or wine," he writes.

For my part, I like to compare BD to yoga. It's a way to strengthen and fortify the whole body, to ward off illness and to maintain health. It helps us sleep at night and relaxes us during the day. It's also a lifelong pursuit, an endless learning process. Any student, no matter how advanced, will discover something new at each class. We've all admired the bodies of those devout practitioners, with their long, supple, and strong limbs. Their frames are so wiry yet flexible that we wonder why they bother practicing so often. Yet they keep coming back for more.

Yoga is self-contained, holistic. There is no court, no playing field, no ball, no bat. No spectators and no competition. You simply use your own body as a tool to shape and strengthen ... your own body. You perform best when you're feeling light and relaxed, so performance-enhancing drugs—even coffee—wouldn't be of any help. Biodynamic farming, too, is (in theory) a closed loop: no inputs from outside the farm should be needed to keep the soil healthy.

There is another, metaphysical, aspect to yoga that isn't much discussed. The same thing goes for biodynamics. It is possible to be a practitioner of yoga without buying into the spiritual side. It is possible to chant *om* purely as a way of exercising the lungs, rather than as a way of achieving that state of mental detachment that allows for meditation. It is possible to be a yogi without being an ascetic or wrapping one's ankles around one's head. And it's possible to perfect each pose while still only getting half-way there, spiritually speaking. Only the yogi knows if she feels a sense of calm and connects with her chakras when she leans into that stretch. Only she knows if it's not just her body but also her soul that is moved by her practice.

Yoga was once the domain of the long haired, the barefoot, and the broke. Now, this spiritual, holistic, and slimming sport has been wrested from the ascetics and adopted by the stars. Hollywood-based yoga guru Bikram Choudhury is said to own a fleet of forty Rolls Royces and Bentleys. Likewise, some of the best wines in the world come from biodynamic vineyards, and fetch hundreds or even thousands of dollars per bottle.

Biodynamic viticulture in Oregon is similar to yoga at your neighborhood studio. Although it's still a fringe phenomenon, it's becoming increasingly popular and voguish. Many winegrowers are dabbling in it. A small number are devout practitioners.

And, yes, there are the very few who embrace the more esoteric elements of it, who speak of the importance of "intention"—a yoga buzzword, too, which infers that mere thoughts can have tangible outcomes—and who claim to see colorful auras around their plants.

But you won't hear much talk about the spirituality of biodynamics among most practicing vignerons. These farmers are more interested in the discipline, and the positive results this discipline appears to elicit. They understand that, if you stretch every day, your limbs will stay supple into old age.

"We are farmers. We are pragmatic, practical people. We don't take a lot of bullshit. You can't just feed us something; we have to experience it," observes Jim Fullmer, executive director of Demeter USA, the biodynamic certification organization. "The roots of biodynamics come from Goethe. It's really simple: Just shut up and observe nature."

Observing nature means going back to the old ways of doing things. It means turning your back on the past century's rapid advances in agricultural science. It means farming like your great-grandparents did, guided by the moon and the stars and aided by the defenses that nature provides.

There is value to a traditional foundation of knowledge. We want a contemporary artist to have learned her trade sketching realistic nude figures; we like it when our favorite novelists make allusions to the literary canon. It's comforting to know that a physician could deliver a baby or resuscitate someone who had collapsed in the street without access to drugs and modern medical equipment.

But biodynamic farmers don't merely rely on a foundation of traditional knowledge; they swear off most modern advances altogether. Or, as one Oregon winegrower so succinctly put it to me, "You really have to know what you're doing. It's like bringing a knife to a gunfight."

Whatever you think about practitioners of biodynamic agriculture, you've got to admit that they've got guts. I don't know about you, but I wouldn't even show up at a gunfight with a gun. Biodynamic farming is like a health regimen of yoga, herbs, and nutrition. Nothing else. Can you imagine living without access to Ibuprofen?

It may be laudable, but it also may be foolhardy. I'm happy to treat my kids' colds with homeopathic syrup, but I also give them antibiotics when they are gravely ill. If I were a farmer, would I be willing to trust the fate of my crop to herbal remedies alone?

How do biodynamic vinegrowers stay safe? They stay on high alert. "What matters is that biodynamic cultivation signals a willingness to pay extreme attention to vines and wines," the wine writer Matt Kramer observes in his recent book, *Matt Kramer on Wine*. "Like driving a race car, if you take your eyes off the road—or in this case, a highly vulnerable vineyard—an irremediable disaster can result. Ask any farmer: attentiveness is always a good thing."

Burgundy's premier domaines farm this way. And in Oregon's Willamette Valley, where the pinot noir grape is everything, all eyes are on Burgundy. Doug Tunnell is considered—by American observers, at least—to be a vigneron in the Burgundian tradition, because his wines are understated and subtle and reek of *terroir*. And because he practices biodynamic agriculture.

If your target demographic is the serious wine geek, there is a marketing advantage to biodynamic certification. Brick House Vineyards is one of only two Oregon wineries belonging to the natural-wine trade group, Return to Terroir, based in France. For lovers of natural wine, for whom "handcrafted," "artisanal," and "authentic" are buzzwords, Doug Tunnell, with his old barn, his biodynamic viticulture, and with straw in his hair, is a cult star.

"Cult" being the operative word: as I was researching this book in 2010, only sixty-eight vineyard properties in the United States could claim to be Demeter Certified Biodynamic®. In Oregon, only sixteen vineyards were certified. This may be a rapidly growing movement, but at the moment, it's still the fringe of the fringe. That said, Demeter's winegrowing membership list had grown nearly fourteen-fold over the previous four years. Every sommelier and wine merchant in the United States was talking about BD, and every serious wine drinker was wondering about it.

And in Oregon, even if they weren't seeking certification, many more vinetenders were dabbling in BD practices. Despite a recession that was debilitating the industry, Oregon winegrowers were relentlessly pursuing quality, an environmentally sustainable form of farming, and a lifestyle that might turn back the clock in the face of the sometimes-terrifying onslaught of technology. They were seeking ways to differentiate themselves and to help each other. And, above all, they were on a quest for that holy grail, *terroir*.

Every toddler mimics the calls of farm animals before learning to speak. Every child grows up familiar with those Platonic images of the big red barn, the white picket fence. Children's books depict a farmer who husbands by hand a wide variety of crops and livestock. Through a child's eyes, the farm is seen as a fertile self-contained ecosystem.

In most of the United States, and, increasingly, the world, that image no longer squares with reality. But biodynamic winegrowers are resisting this contemporary irony. They believe it's time to get back to basics, to the kind of farming we learned about in preschool.

A is for animals, which are an essential part of any working farm. Bovine manure, deer bladders, weed-eating sheep, and gopher-killing kestrels are all part of a successful biodynamic ecosystem.

B is for Burgundy, and for biodynamic. In Oregon, the biodynamic Bs include three of the most revered producers in the state: Bergström, Beaux Frères, and Brick House, where Doug Tunnell labors, with straw in his hair, and worms in his hands, and almost—almost—Burgundian wines in his barrels.

C is for the cosmos and for cow horns, those otherworldly aspects of biodynamics that fascinate some observers and repel others. It's also for centuries and centuries of farming this way, and for the past century, when we forgot how it was done.

CHAPTER ONE
In the Old Country

"Quien trabaja con plantas, las conoce. Pero ¿y si no la conoces?
Entonces debes saber cómo sentir su fuerza vital."
—Luis Alberto Urrea, *La Hija de la Chuparrosa**

In September 1982, Moe Momtazi sat on a motorcycle with his heart in his throat and his hands gripping the waist of a drug runner. A few possessions—clothes, some food—were stowed on another bike. On a third, her swollen belly pressed against the back of another smuggler, was his beautiful wife, Flora, eight months pregnant with their first child.

The Russian motorcycles were built for tough terrain, but in places, they were no match for the rocky footpaths climbing the desolate Siahan (or "Black") Range. When the trail became too narrow or the broken shale too jagged under the tires, the riders would dismount and walk. At times, when their speed felt reckless, Moe would feign clumsiness and drop the green bag he clutched, so their guides would be forced to stop and Flora could rest a little.

When they finally crossed the border into Pakistan, Moe's driver whooped with excitement and accelerated, promptly speeding blindly off a ten-foot embankment. The moment the bike had landed—safely, in sand—Moe turned and screamed for Flora's driver to stop. His shouts weren't heard over the engine's roar. Flora flew off her motorcycle and into the sand, dazed. The baby stopped kicking.

This journey had not been part of the plan for Moe Momtazi, a bright and ambitious young Iranian who had recently founded his own firm after studying civil engineering in the United States and working for a prominent construction company in Tehran and Zanjan.

But fate and politics had intervened. Three years previously, the Ayatollah Ruhollah Khomeini had deposed Shah Mohammad Reza Pahlavi. Then, from November 4, 1979, to January 20, 1981, a group

---

* In English: "To work with plants, you must know plants. But what if you don't already know a plant? Then you must know how to feel its life force." —*The Hummingbird's Daughter*

of Islamic militants had held fifty-three Americans hostage, bringing about American economic sanctions that had crippled the nation's economy.

Suddenly, the new revolutionary government teetered on the verge of bankruptcy, unable to pay Momtazi for construction projects his firm had completed. His business prospects dimmed as he fruitlessly tried to reason with newly appointed fundamentalist bureaucrats. "The sector of the government I was dealing with, telecommunication, was all run by twenty-something-year-old kids, fanatics, who had no idea what they were talking about. They couldn't make any sense of a blueprint," he recalls.

Meanwhile, the proud Persian society he had known since childhood was crumbling all around him. Friends and cousins were being imprisoned and executed by the state. If Moe and Flora were going to have any future, they had to get out of Iran.

The opportunity to flee came up when Moe Momtazi's brother Ahmad, an orthopedic surgeon, was serving his medical residency in the remote eastern region of Baluchestan. Ahmad operated on a crippled boy, enabling his patient to walk. Afterward, the child's family thanked him in a cryptic manner. "The relatives of this kid came to my brother and said if he needed anything, they would take care of it. Then he found out that they were drug runners, who would smuggle not only drugs but human beings," Momtazi recalls.

Despite the tales they had heard about families disappearing during dangerous border crossings, Moe and Flora decided to take up this opportunity to escape. Moe tucked approximately $3,000 of travel money into a simple green woven bag and climbed on the back of a motorcycle, placing the fate of himself, his wife, and his unborn child in the hands of a gang of drug smugglers.

Once across the Pakistan border, Moe climbed into the back of a tiny compact pickup truck, a space he—impossibly, it seemed—shared with fifteen other passengers. Flora sat silently in the cab. They stopped in a village, where a midwife examined her. "I think your baby is going to be OK," she tentatively assured the terrified young mother.

Four days after they had first set out from Iran, the truck pulled into the city of Quetta. Moe reached into his green bag and paid each of the drivers, then took Flora to a hospital. The resident obstetrician examined her and declared that he didn't think the baby had been harmed by the tumble off the motorcycle. The two young parents heaved a sigh of relief. But their ordeal wasn't over.

The couple secured a visa to fly to Spain, where Flora could deliver her baby, a miraculously healthy girl. From Spain they moved on to Italy. From Italy they traveled back through Spain to Mexico. Finally, when their daughter was three months old, they entered the United States. They were alive and healthy, but by this point, nearly penniless. It was January 1983.

Moe felt comfortable here, having studied civil engineering at the University of Texas at Arlington. He quickly obtained a work permit, then a green card. He began designing trusses for firms in Dallas and Austin. Five years later, he achieved citizenship. He established and sold a truss-production company in Atlanta. He was back in the business of pursuing his dreams.

Then he started looking for a new, fast-growing market where another truss manufacturer would thrive. He and Flora settled upon Oregon. They lived first in the town of Sandy at the base of Mt. Hood. In 1994, they moved their then-four-year-old business, Tecna Industries, to McMinnville, in the Willamette Valley. The couple by now had three children, a trio of schoolgirls, and Flora was working at Tecna alongside her husband. Iran was becoming a distant memory, a faraway place where relatives still picked up the phone. Their life was here, in the United States.

But as Moe explored the lush landscape surrounding McMinnville, his thoughts traveled back in time. He began to think about the sweet childhood summers he had spent on his grandfather's farm near Lahijan in northwestern Iran. That property, blessed with fresh breezes off the Caspian Sea and fertile soil, included a tea plantation, orchards, and rice paddies. Oregon's Willamette Valley, cooled by winds off the Pacific Ocean, is similarly fertile, at least along the valley floor, which is known as the "grass-seed capital of the world."

But the hillsides are different. Soaked in summer sunshine and cool-season rain but poor in soil nutrients, they harbor hazelnut groves and—more interesting to the Momtazis—vineyards. Here, touring and tasting at the wineries around McMinnville, Moe and Flora fell in love with the heartbreakingly beautiful pinot noirs of the region.

Although alcohol had been forbidden by the time they left Iran, the couple had fond memories of wine from before the revolution. Persia had a winemaking tradition dating back as far as seventy-five hundred years ago and a thriving fine-wine industry during Moe's and Flora's youths. Moe's father had always made wine at home to share with his family, and even named his son after his hobby: the Persian definition of the word "moe" is grape leaf, or grape vine.

In 1997, Moe and Flora decided to give grape growing a try and bought an abandoned wheat farm south of McMinnville. At nearly five hundred acres,* the property was huge by Oregon vineyard standards. Ranging in elevation from 192 to 783 feet, it didn't seem to have a single square foot of flat ground. But Moe Momtazi, the structural engineer, was undaunted. He dug three massive irrigation reservoirs and stocked them with trout, to act as biological monitors of the water's purity. He built a small sawmill where he could transform the property's timber into lumber. He designed and built a winery, a greenhouse, and other structures, using materials from the property and his own know-how. He built a winery and named it Maysara, a Farsi term meaning "house of wine."

When it came time to plant and tend the grapevines, Moe thought about his grandfather's farm. Although neighboring landowners back in Iran had begun to embrace American-produced chemical fertilizers and pesticides as they became available, Momtazi's grandfather had always spurned these additives, preferring to practice a traditional, self-sufficient form of agriculture he called "natural farming."

"We thought that he was not right, because whoever was using chemicals was producing more," Momtazi recalls now. "But his main argument was that food should not be just a stomach filler. And in order to achieve purity in your food, there should not be any manmade product in it. Of course as I age, I realize now that he was right. Some of the chemical products do produce a lot more food, but with very little flavor and not much nutrition."

To honor the memory of his grandfather, Momtazi began to farm his new vineyards organically. Then, along with his energetic young winemaker and vineyard manager, Jimi Brooks, he joined a study group made up of other Willamette Valley winegrowers.

---

* The Momtazi Vineyard estate has since expanded to encompass 532 acres.

The topic was an obscure form of sustainable farming called biodynamic agriculture. The teacher, Andrew Lorand, was a "real biodynamicist," as one former study-group member now wryly puts it. Lorand's workshops were peppered with activities such as drawing and molding clay. He talked about seeing colorful "auras" in the vineyards. He assigned an esoteric text by Rudolf Steiner, the founder of biodynamics, entitled *How to Know Higher Worlds*; and he led discussions about the unfolding of the chakras and the spirit world.

In short, the man was a walking billboard for everything that the general public thinks is completely daft about the biodynamic school of thought. His students were circumspect, taking Lorand's words with a few good grains of salt. But as Lorand rambled on about Steiner's religious writings—which deemed Jesus Christ a reincarnation of the Persian prophet Zarathustra and described a destructive figure known as Ahriman—and the other students rolled their eyes, Moe Momtazi's mind was racing. A lot of this talk sounded familiar to him. And in the context of his personal history, it made some sense.

Indeed, Rudolf Steiner had lifted large sections of his theology from Zoroastrianism, a monotheistic religion more ancient than Christianity and practiced by the Parsis of India. "Steiner himself was honest enough that he never really claimed that he created this," Momtazi points out. The spiritual home of Zoroastrianism is Persia, but practitioners there have been persecuted for the past thirteen hundred years. Like many Persians, Momtazi's great-grandfather had been born a Zoroastrian, but had converted to Islam as a young man. But while the outward practice of this religion had faded over time, the folklore of the ancient creed lived on among the Persian people, especially in the countryside, where, in the Zoroastrian tradition, farmers consulted the heavens before sowing and reaping.

The herbal preparations that biodynamic farmers sprayed onto their crops and integrated into their compost were familiar to Momtazi, as well; the very same herbs were used back in Iran to make homeopathic remedies used to treat health problems.

The more Momtazi studied biodynamics, the more he realized it was a style of agriculture that could be traced back to his childhood and, indeed, the earliest Zoroastrians. Many aspects of its creed that sounded odd to Western ears sounded familiar to Middle Eastern ones. Biodynamics was just another term for the same "natural farming" his grandfather had practiced.

His memories of that idyllic farm were so distant now. In 1963, under pressure from the Kennedy administration to, as one writer puts it, "resist the so-called menace of communist encroachment," the Shah's government had confiscated large estates such as Moe Momtazi's grandfather's farm. Dubbed the "White Revolution," the mass seizure had been a foreshadowing of what was to come.

Momtazi had been a boy of twelve in 1963. He had come so far since then. Today he was a successful American businessman with a healthy family and a new vineyard. But he still kept a frayed green bag that he looked at from time to time, "To bring me back to earth, to not forget all that has happened."

To his classmates, biodynamics were a leap of faith. To Moe Momtazi, biodynamic farming was a way of returning home.

The fundamentals of biodynamic agriculture are these: Time farming decisions according to the movements of the moon and stars in the sky. Use the raw materials on your property to nourish your crops. Protect nature, which in return will protect your harvest. And, in doing all these things, harness the spiritual forces of the heavens.

Some of these maxims might sound unusual to us, but the fact is that humans have been raising crops this way since the dawn of civilization.

Blame it on the moon.

In the right hand of the *Venus with Horn* of Laussel*—a Paleolithic fertility figure carved in limestone and painted with red ochre—is a crescent-shaped bison horn with thirteen ticks in it, which experts believe to be a prehistoric lunar calendar. She looks like she could be the twenty-five-thousand-year-old patron saint of biodynamic agriculture, holding her cow horn aloft.

Let's fast-forward through time. Archaeologists place the first use of a harvesting tool, a crescent-moon-shaped sickle, at around 11,500 B.C.E. It was used by the Natufians, a Mesolithic people living in the Levant, not far from the place where, some six thousand years later, the Mesopotamians would establish the world's first civilization.

---

* Now in Le Musée d'Aquitaine, in Bordeaux.

While they worshipped the moon as a deity, the Mesopotamians also observed it analytically, creating a twelve-month lunar calendar very much like the one we use today. For them, astronomy and astrology—science and spirituality—were one and the same. As far back as the sixth millennium B.C.E., Mesopotamian farmers consulted the constellations to determine the best time to plant their crops; and by the third millennium, they had assigned these constellations the same symbols of the zodiac that we today see when we check our horoscopes.

Historians mark the end of Mesopotamian civilization at 539 B.C.E., when the by-then Assyrian-ruled empire was overthrown by Cyrus the Great, a Zoroastrian and enlightened Persian leader who instituted policies of religious freedom. For two centuries, the traditions of the Mesopotamians and the folklore of the Zoroastrians mingled; then, Cyrus's empire fell under the attack of Alexander the Great in 330 B.C.E.

The Greeks of antiquity embraced the late Mesopotamian notion of astral determinism: Plato writes that events are proscribed by stellar and planetary conditions and—as the biodynamic wine expert Monty Waldin has pointed out—Hesiod liked to dabble in oenological advice, specifying that winemakers time their vinification activities according to the positions of Orion and other constellations.

We can also, perhaps (or perhaps not), thank the Greeks for *Secretum Secretorum*, a purported letter from Aristotle to Alexander the Great that appeared in Arabic sometime around 941 A.D. This "Secret of Secrets" treatise was really more of an encyclopedia, covering topics ranging from diet and health to the conduct of war. "Do not rise or sit or eat or do anything except at the time chosen by astrology," warns the apocryphal author, before providing a detailed guide on the properties of various curative herbs and their relationships with the planets. It all sounds quaint today, but for some six hundred years the *Secretum Secretorum* was one of the most widely read—if not *the* most widely read—text in all of Europe.

Continuing a tradition seven millennia old, medieval European farmers considered an understanding of the heavens elemental to an understanding of plant growth, and, indeed, life on this earth. Early clocks of the Middle Ages typically displayed the phase of the moon and the position of the sun in the zodiac as prominently as the hour and the minute. As the Bolognese grapevine expert Pietro Crescenzi wrote in his *Ruralia Commoda,* "the moon could do in a month what the sun does in a year."

Crescenzi's work sat dormant for more than one hundred fifty years thanks to the Black Death, which quickly wiped out between 30 and 60 percent of the European population in the latter half of the fourteenth century; written in 1305, it was only published, post-pestilence and post-Gutenberg, in 1471. During this time, the demand for food—and thus the impetus to document and improve farming techniques—wasted away like a victim of the bubonic plague.

It wasn't until the sixteenth century that the population surged again. Unfortunately, the fine art of farming had been neglected for two centuries by that point, and agriculture couldn't keep pace with the need for sustenance. "Crop failures might occur as often as every dozen years. A drought or excessive rainfall could ruin an entire season's work. Wars could completely dislocate the rural economy," writes historian Ken Albala. The Renaissance men of the late sixteenth century rose to the challenge, publishing a flurry of new agricultural references: "a new attitude toward farming and a flourishing literary genre that catered to this growing interest in agriculture as a worthy pursuit for the leisured nobleman," Albala writes.

For example: Thomas Tusser, an Eton and Cambridge alum who published his definitive work, *Some of the Five Hundred Points of Good Husbandry,* in 1580. Like any modern-day biodynamicist, Tusser was a strong proponent of compost (or *compas,* as he spelled it) and instructed his readers to "Sow peason and beans in the wane of the moon, / Who soweth them sooner, he soweth too soon."

By the eighteenth century, forward-thinking British agriculturists were analyzing and updating the age-old folkways. With study and testing, cover crops, rotation, composting, and manure-based fertilizers brought about not merely survival, but profitability. (Of course, it helped that British inventors were simultaneously developing the mechanical seeder and the threshing machine.)

As the advancements of the Agricultural, and then the Industrial Revolutions spread from the United Kingdom to the rest of Europe and the United States, laborers on both sides of the Atlantic moved from the countryside to the cities, from farm work to factory jobs. Farms became large-scale operations, which necessitated the importation of guano (the excrement of tropical sea birds, high in phosphorus and nitrogen) to maintain the fertility of the soil. When the guano ran out, American fertilizer companies harvested deposits of sodium nitrate from Chile.

The First World War marked the beginning of the end of the old ways. "The appearance of a few drops of colorless liquid at one end of an elaborate apparatus in a laboratory in Karlsruhe, Germany, on a July afternoon in 1909 ... was to have arguably the greatest impact on mankind during the twentieth century," writes the journalist Tom Standage. "The liquid was ammonia, and the tabletop equipment had synthesized it from its constituent elements, hydrogen and nitrogen. This showed for the first time that the production of ammonia could be performed on a large scale, opening up a valuable and much-needed new source of fertilizer and making possible a vast expansion of the food supply." With these new chemicals, recently perfected farming practices such as cover crops, rotation, and composting suddenly became irrelevant. According to historian Mauro Ambrosoli, "The scheme for the constant improvement of the soil to support crop-growing was largely abandoned."

A 1916 American reference entitled *Successful Farming* is heavy on what we would now call organic and biodynamic practices, but the information is couched in the cause-and-effect language of modern science rather than that of homespun folk wisdom. Crop rotation is advocated to prevent disease; cover crops are encouraged as a source of increased fertility; homemade manure-based compost is promoted as the best and least expensive soil amendment; and the pasturing of animals is praised as a way to improve soil conditions while providing feed.

Yet amidst all of this is a chapter on the subject of commercial fertilizers: "A careful study of the condition of farming in the United States shows that the supply of barnyard and stable manure is not adequate to maintain the fertility of the soil. The need for commercial fertilizers is, therefore, apparent and real," writes the author, Frank D. Gardner. "This is but natural, since there is a constant flow of soil fertility towards the cities. The rapid increase in the city population and the consequent increase in food consumption at those points cause a constantly increasing drain upon the soil fertility of the farms."

As is so often the case with history, Rudolf Steiner, the father of biodynamics, was the right guy in the right place at the right time. He grew up observing the old ways of the peasant farmers of late-nineteenth-century central Europe. He was swept up in an early-twentieth-century craze for the study of ancient religions such as Zoroastrianism. And he happened to be lecturing and writing just after the First World War, when weapons factories were producing the earliest chemical fertilizers and herbicides.

The agricultural philosophy that Steiner outlined in his series of 1924 lectures exhorted farmers to turn their backs on the siren call of the easy new chemicals. But by then, it was too late for most; they had grown accustomed to the ease of dumping fertilizer on their soil, so why should they go to the trouble of composting?

The great irony is that today, nearly a century later, our agricultural system, now reliant on chemical supplements, suffers from an excess of excrement. In what would appear to our forefathers to be a ridiculous problem—Frank D. Gardner's chief concern in 1916, remember, was that "the supply of barnyard and stable manure is not adequate"—factory farms produce such an abundance of waste that it has become a public health hazard.*

But the disconnect isn't universal. A small but increasing number of farmers in the U.S. has rediscovered the modest, self-sustaining farming styles that marked the onset of the Agricultural Revolution. They're saving seeds and nurturing heirloom vegetables. They're composting, growing cover crops, and raising a diverse array of produce and animals rather than following the agro-industrial monoculture model. They're fueling their farms with animal poop. They may call themselves "sustainable," or proudly display their USDA Organic seal. They may even time their agricultural activities by the phases of the moon and have biodynamic certification. But in essence, it's all the same idea: chemical-free farming.

---

* Even if all of that excrement were to be composted and applied to crops, the amount would still, surely, be excessive. According to the USDA, American per-capita consumption of poultry increased 450 percent between 1910 and 2000, while per-capita consumption of red meat increased 19 percent. The bottom line: We consume—and, therefore, are producing—double the federal government's recommended per-capita allowance of protein daily.

"When I was a little boy, I remember my grandpa waiting for the moon cycle to start planting or harvesting. It was the same with my dad," recalls Juan Pablo Valot, vineyard manager at Silvan Ridge and Hinman Vineyards in Eugene. "Then, when I went to college, we were told to forget about the old-fashioned ways. It was all chemicals, fertilizers, fungicides. It was all about new technologies and more input."

Valot is Argentinean and teaches Spanish-language viticulture courses to Latino farm workers through Chemeketa Community College's Northwest Viticulture Center in Salem, Oregon. Although folk-farming traditions in Mexico look very much like what we would call biodynamic agriculture, Valot says that his students didn't grow up learning those traditions. According to the National Center for Farmworker Health, Inc., half of all immigrant farm laborers in this country are under the age of thirty-one, and most come from urbanized areas. "All the Mexicans that are working here, they learned what they know about farming here in the U.S.," says Valot.

Scott Shull, winemaker and owner, with his wife Annie, of the Willamette Valley winery Raptor Ridge, has remarkably similar childhood memories to Valot's, even though his family lived thousands of miles away from Argentina, in the American Midwest. His uncles were among the first to adopt contour farming to combat erosion in the Kansas Dust Bowl. "My uncles did things by the *Farmer's Almanac*, yet at the same time they were some of the most advanced farmers of their day," Shull recalls. "They planted their root crops by the dark of the moon, and their vegetables and grains by the full moon. They used the manure from their animals to fertilize their farms. My mom sent me a picture of their four-horse team and their manure spreader. They would take the output from the dairy cattle and put that on the wheat fields."

Like Moe Momtazi, Shull studied and practiced engineering prior to coming to winemaking. Unlike Momtazi, he has trouble squaring his scientific training with the quasi-religious overtones of biodynamic agriculture. At the same time, his knowledge of the direct link between weapons manufacture and the agrichemical industry makes him queasy. (As the journalist Tom Standage has noted, the Nobel-prize-winning German chemist Fritz Haber spent the First World War developing both chemical fertilizers *and* chemical weapons: "The fact remains that the man who made possible a dramatic expansion of the food supply, and of the world population, is also remembered today as one of the fathers of chemical warfare.")

"There was a consistent pattern, especially in the 1940s and '50s, of conventional farmers using heavy agrichemical inputs and coming to very much regret that," says Shull. "The poster child is the weed species that have become resistant to Roundup, whereas there is not a weed I know that has become resistant to a hoe."

Shull follows the moon cycles in the vineyard and the winery, in deference to his Dust Bowl predecessors. And he farms sustainably simply because it makes sense: "I looked at buying commercial fertilizers, and I looked at planting cover crops and tilling, and guess which was cheaper? By a factor of about ten, it was cheaper to plant seeds and till, even if it takes a little longer. So is that conventional? Is that organic? I think it's just smart."

In many ways, winegrowers like Shull are turning back the clock to, say, 1917: a time when mechanization had made farming manageable,* but chemical additives weren't yet ubiquitous. And every time they consult a lunar chart, they're commemorating their forefathers.

Idealistic? Unrealistic? Perhaps. But when you consider that, over the past couple of decades, an ever-increasing number of American palates have reverted to favor old-fashioned handmade foods over high-tech pasteurized ones, a return to our grandparents' way of growing grapes may not be misguided.

Razvan Andreescu was born in southern Romania, and yet his story is strikingly similar to Moe Momtazi's. Some twenty-five hundred years ago, Andreescu's and Momtazi's homes were both part of the massive Persian empire—stretching from modern-day Libya east to the Indus River, and as far north as the Aral Sea—that the Zoroastrian King Darius the Great built upon the foundations laid by his forebear, Cyrus.

Andreescu was born in 1945, in the village of Turnu Măgurele, cradled by the Olt and Danube rivers and about three miles from his family's historic estate. He was an only child but came from a

* Henry Ford began manufacturing his Fordson tractors in 1917.

large family that had, on his mother's side, grown grapes and made wine for more than three centuries. But young Razvan never got the chance to taste the fruit of those vineyards, however, because they were confiscated by Soviet troops in 1944. Instead, the boy spent his summers with his grandmother, Iona Iordan, on the scant two acres she was left with after the bulk of her property had been appropriated. Like the Momtazi family, the Andreescu family continued to make wine at home, for personal use, but their centuries-old oenological legacy had been broken.

Andreescu is white-haired now and wears crisp titanium-framed spectacles. There is a melancholy quality to his speech, which is richly accented with the cadence of central Europe. As he sits in the cellar of the magnificent winery that he designed to look like a Romanian fortress, his voice cracks and softens a bit as he explains how, in 1948, the newly appointed secret police of the communist regime arrested his father, Virgil, a former member of the Royal Security Police. And how his mother followed her husband to Bucharest to find work near the prison, leaving her son in the care of his widowed grandmother.

If it was a lonely childhood, it wasn't without its idyllic moments. After school, young Razvan liked to visit a small clearing in the woods next to a pond teeming with ducks and other birds. "I used to just lie down and look at the sky through the trees and listen to the wind and all the sounds of the forest. It was beautiful," he recalls.

After eight years of captivity, Virgil Andreescu was released. Razvan joined his family in Bucharest, where he attended university. After serving his compulsory service with the army, the young man returned to Bucharest to settle into a career as a civil engineer. But like Moe Momtazi and Scott Shull, Razvan Andreescu is an engineer with an entrepreneurial streak. After the 1989 Romanian revolution and establishment of democracy, he quickly took advantage of the new freedoms afforded by capitalism and founded a company, an exporter and importer of rolling stock and locomotives.

Then, when their children enrolled in university in Florida in 1991, Razvan and his wife Felicia moved to Miami to be near them. While maintaining part-ownership of his company in Bucharest, Razvan began working in commercial and residential development and construction in the United States. By 2005, he was ready to retire and realize his lifelong dream: to establish a vineyard and winery, and reestablish his family's legacy. He and Felicia began to travel west, with the thought of ending their journey in California. But in Oregon, as they drove west along the Columbia Gorge, they gasped. "I saw this big river and it looked so much like the Danube that I said, 'This is the place I want to stay,'" Razvan recalls.

The couple began looking at properties in the Willamette Valley and learned that the revered Napa- and Willamette-Valley vintner Tony Soter was getting ready to build a winery on his recently acquired Mineral Springs Ranch. In order to make the move, Soter sold his Beacon Hill Vineyard in the Yamhill-Carlton District appellation to the Andreescus, purchasing the fruit back from them through the 2006 vintage. This gave Razvan and Felicia time to think about how they could improve the site.

Razvan learned that some of the top estates in the Willamette Valley, such as Beaux Frères, were working with a French biodynamic consultant named Philippe Armenier. His engineer's training had instilled in him a sense of respect for efficient systems; this made him eager to try out this reportedly self-sufficient form of farming. He began reading Rudolf Steiner's lectures and hired Armenier to teach him how to plant the herbs that could be used to make biodynamic preparations, as well as how to build compost piles, make teas, and treat the vineyards.

It quickly became apparent that Beacon Hill Estate was the biodynamic ideal: located at a sleepy conjunction of country roads, near the hamlets of Gaston and Yamhill, it was protected on three sides by forest; its wildflowers and oak trees attracted wildlife from the surrounding woods. Owls, hawks, and bobcats hunted the gophers that threatened the vine rows; deer drank from a freshwater spring that could also be used for washing barrels and equipment.

Razvan was struck by how similar this environment was to that of his grandmother's farm in Romania, a nation notable for its undisturbed evergreen forests, oak groves, lynxes, and birds of prey. Add to that the techniques he was learning from Armenier and he felt like he had traveled sixty years back in time: "Believe me, I didn't pay any attention at the time, but I now remember," Andreescu says. He remembers the stories he heard as a child about his grandfather's peculiar farming practices that mixed modern insights with age-old wisdom—consulting the lunar calendar before making planting and harvesting decisions, for example.

In the Old Country

He remembers watching his grandmother carefully preserve heirloom vegetable seeds. And how, as he played in a field nearby, she and some farmhands used to cover the bare earth with compost before the long, cold winter, then plow open the soil in the spring. "You would grab a fistful of soil and smell it and sometimes even taste it. This comes naturally. You don't have to be trained in it," says Razvan, who now lives on Beacon Hill Estate with Felicia. His ninety-year-old mother, Ana, lives with them part of the year, too; she goes out and works in the vineyard every day. "She remembers," Razvan chuckles. "She remembers what she saw her father doing. She knows how to prune."

When I visit the Andreescus nine months after our initial meeting, they've been through a rough patch. The global recession hit their company back in Romania hard; their fledgling wine business here in the U.S. struggled against the competition. At the economy's nadir, in the early part of 2010, they reluctantly put the estate on the market.

Throughout this jittery year, even as they struggled to make ends meet, the Andreescus never gave up on biodynamic farming. Philippe Armenier continued to visit, refusing to charge them for his time.

"I wouldn't have done it any differently," Razvan says. "When I got here, there were no grasshoppers, no dragonflies. No bugs. No worms in the soil. Now the soil is full of worms." He holds his hands to his nose and releases them. "And oh, the smell. It must be happy." The estate does seem … happy. It's alive with birdsong and buzzing insects. Despite the fact that it isn't irrigated, the vineyard glows, a saturated shade of green. Even the weeds have a sort of luminous beauty: Once hard and intractable, they're now soft and pliant. They smell like fragrant herbs when pulled from the earth.

Surrounded by nature is Razvan's tour de force and possible folly: the massive old-world-style winery. The second time I visit, it's still incomplete, with plastic sheeting over the windows. But as Andreescu walks around and shows off the thick French doors, the massive stone fireplace, and the picturesque balconies, it's easy to see how an engineer-cum-real-estate-developer couldn't resist such a project. The man is a builder. He had to build.

And he is finishing the building. He has found other winemakers who have contracted to buy his biodynamically farmed fruit. The economy is picking up. His business is doing better. He has taken the estate off the market.

I ask Razvan Andreescu if there's any irony in the fact that someone who made his fortune in real estate development and heavy equipment sales, a man who felt compelled to build a gargantuan winery, should embrace a holistic, hands-on, back-to-the-land form of farming. "You don't have to destroy Mother Nature to create habitat for humankind," he responds. "I did develop shopping malls and buildings like condominiums, but always in a location that was previously used for the same purpose."

But still. I'm having trouble squaring the passion that men of science, mathematics, and business, like Moe Momtazi and Razvan Andreescu, have for what appears to be a mystical form of farming. Sure, it hearkens back to the old-world ways followed by their forefathers, but both men intentionally left the Old World behind them when they moved to the United States. Scott Shull is a man of science who appreciates the old ways, but he doesn't dwell in the past.

"I think biodynamics is the future," Razvan Andreescu retorts. "And no matter how clever you are, how intelligent, using chemicals and high-efficiency man-made products just to make it easier is not the key to achieving the real work of Mother Nature." He speaks slowly and deliberately. "Just supervising or fine-tuning the work of Mother Nature will allow you the highest potential of the respective crop. We should promote this for the benefit of humanity. It's healthy."

When we finish talking, I ask Razvan Andreescu if he knows Moe Momtazi. He doesn't. I tell him I think they might like one another; that they have a lot in common. And after I leave, I realize I forgot to mention the other person Razvan and Felicia—and their children, too—might like to meet: a pretty young woman who traveled all over the world to learn her craft before becoming the youngest female winemaker in the United States.

She was born in Spain after a rough ride through Persia and Pakistan. Her name is Tahmiene Momtazi. She is the winemaker at Maysara.

CHAPTER TWO

The Gospel of Rudolf Steiner

> And Noah began to be an husbandman, and he planted a
> vineyard. And he drank of the wine ...
>
> —*Genesis*, 9:20-21

Rudolf Steiner was the most remarkable scholar you've never heard of.
This freakishly accomplished polymath wrote twenty-five books and
delivered more than six thousand lectures on at least three hundred
and fifty different subjects during the course of his lifetime. And he
wasn't merely voluminous. He was a truly original thinker, whose
keen insights into the realms of education, medicine, philosophy,
science, art, drama, literature, architecture, and agriculture still
guide practitioners of all of these disciplines today. His overarching
philosophy, anthroposophy, has inspired a worldwide spiritual
movement.

But despite the man's myriad accomplishments, Steiner hasn't
made it into any mainstream history textbooks. Because his encyclopedic knowledge and keen powers of perception haven't been half
so interesting to us, in hindsight, as his musings on the sunken city
of Atlantis, two characters by the name of Ahriman and Lucifer, and
the importance of cosmic destiny.

In short, the biggest stumbling block for those of us trying to
understand biodynamics is the esoteric mind of the man who codified
this compelling—if also somewhat surreal—school of agriculture.

Plenty of practitioners of biodynamics say we shouldn't heed
Steiner's more colorful beliefs. But is it fair to the movement or
the man to ignore these beliefs and pretend that they don't exist—
to proselytize the message while spurning its messenger? Many
parents of Waldorf School students would say so. Families who find
the arts-oriented environment of these Steiner-influenced academies
to be ideal for their child's style of learning might still have serious
qualms with the quasi-Christian, occultist leanings of some of the
teachers.

And if we were to delete from the canon the work of every
historically significant thinker whose views we vigorously disagree
with, college reading lists would grow woefully slim. Immanuel

Kant and T. S. Eliot? Anti-Semites. Friedrich Nietzsche? Misogynist. And so on. Slave owners, racists, homophobes … they've all written books full of important ideas that the most open minded of us embrace today.

And Steiner's off-the-wall notions weren't overtly malicious.* In 1923, while Hitler was busy composing *Mein Kampf* and hatching his plans to eliminate—among many others—the disabled, Steiner was busy hatching his plans for the curative education of the severely disabled. As is so often the case with history, the crazy guy with the terrible idea is the one who has enjoyed worldwide infamy. The crazy guy with the good idea teeters on the verge of oblivion.

I think more of us might be interested in Steiner if his biographers hadn't all been so *earnest*. Most accounts of the man's life were scribbled by starry-eyed disciples who have nothing but banal things to say about the man. The only even remotely meaty tome on the subject of Steiner's life was written, refreshingly, by the occult expert Gary Lachman, who was one of the founding members of the punk band Blondie.

Reading Lachman's account, one sees Steiner as a Zelig-like figure, a chameleon able to ingratiate himself into any milieu. Here he is, donning a foppish scarf and debating politics with Austrian intellectuals at a Vienna coffeehouse. There he is, lecturing to disgruntled laborers one day, then hobnobbing at a literary salon the next. He's sermonizing about the virtues of Jesus Christ and the Buddha, then—HELLO!—hanging out with a group dedicated to magic-sex rituals (despite the fact that scholars aren't sure whether he ever lost his virginity).

This should be scintillating reading. This could have been a Stanley Kubrick film. Instead, Steiner's biographers treat all the juicy stuff as mere asides: "Oh, you know that clever man who founded the Waldorf education system, invented anthroposophic medicine, and conceptualized biodynamic agriculture? He also liked to talk to dead people."

So let's not make Steiner an afterthought in this exploration of biodynamic viticulture. Let's face him head-on. But as we examine

---

* That said, they could be extremely ill informed and unpleasant, such as when he made statements about "ascending" levels of race and described Africans as passive and childlike.

his eccentricities, let's give them some historical context. Because as one studies Steiner's upbringing and milieu, one sees a method to his madness. If he was already intrinsically a bit fey as a child, the circumstances of his life surely drove him even further in that direction.

Rudolf Steiner was born in 1861, shortly after the Second Industrial Revolution had turned the career path of his father, Johann, 180 degrees, from estate gamekeeper to telegraph operator with the Austrian Southern Railway. Likewise, young Steiner's earliest memories must have been bifurcated between the thrum of industry and the primal peace afforded by nature.

During Rudolf's childhood, the family moved frequently throughout Central Europe as his father's career developed. The lonely boy sought solace in the outdoors, lolling in the fields and contemplating the old ways of the local peasants. He was fascinated by the bucolic existence they eked out from the land, guided in their work by folklore, intuition, and the positions of the stars in the heavens.

At the same time, young Rudolf learned some of his first school lessons by his father's side in a train station, watching the massive machines come and go. And for entertainment, the boy liked to visit a nearby mill and textile factory, where he would observe the humming progress of the equipment inside. Torn between the opposing forces of the pastoral and the industrial, the spiritual and the sensual, Steiner developed a lifelong sense of cognitive dissonance. It was a popular problem of Germanic thought at the time; eighty years prior, Johann Christoph Friedrich von Schiller had written about the friction between man's sensuous side and his rational side, and his ideas were still a topic of public discussion and debate.

Steiner eventually worked out his inner conflict by imagining that a couple of amoral devils reside in every human soul. "Lucifer"*

---

* The name, from the Latin, meaning "light-bearer," or Morning Star, does not refer to Satan. "One's childhood picture of Lucifer as a slithering manifestation of evil is difficult to reconcile with the beauty of this name," lectured Steiner. "Lucifer, however, represents a force that paradoxically can combine beauty and, if you will, beauty gone too far, to the extreme of decadence, hence to evil." To further confuse matters, Steiner referred to Lucifer as "the devil" and Ahriman as "Satan."

was a person's spiritual side, in touch with nature and creativity yet out of touch with reality, while "Ahriman" was the mathematical, rational self, favoring information over inspiration. A person too susceptible to Lucifer might live in a world of delusion and fantasy; someone ruled by Ahriman might be materialistic. As with yin and yang or the components of a fine wine, the trick was to keep these opposing elements in balance with one another, and their worst influences in check. This search for stability would be an ongoing theme in Steiner's life whether in the arenas of culture, metaphysics, or politics.

Another lifelong theme for Steiner was cultural diversity. The son of two German-speaking Austrians, he was, as a child, cut off in both language and custom from his Hungarian and Slav neighbors in what is now known as Croatia. But the Austro-Hungarian Empire of the day was a rich cultural patchwork of Bosnians, Croats, Czechs, Germans, Hungarians, Italians, Poles, Romanians, Ruthenians, Serbs, Slovaks, and Slovenes. And so the boy raised in train stations, watching people of all walks of life come and go, grew up to be a chameleon-like citizen of the world, comfortable lecturing in foreign nations and connecting to a wide range of peoples.

"Although decidedly Germanic," writes Lachman, "Steiner later believed that the mix of cultures and nationalities that surrounded his early life primed him for part of his spiritual mission: to act as a kind of meeting ground between the mysticism of the East and the materialism of the West."

The flip side of such a culturally rich childhood was the post-revolution rancor in the unstable Austro-Hungarian Empire, and an early adulthood spent in an increasingly nationalistic German Empire. It can't have been easy to grow up in the environment that bred the First World War and then Nazism. Steiner dealt with the instability around him by believeing that he possessed supernatural powers that would help him to rise above the chaos.

"We don't like uncertainty, which can be psychologically corrosive," observes noted British psychologist Bruce Hood,[*] author of the recent book *Supersense: Why We Believe in the Unbelievable.* "So

---

[*] Hood is Director of the Bristol Cognitive Development Centre in the Experimental Psychology Department at the University of Bristol, United Kingdom.

we tend to engage in beliefs and behaviors which give the perception of control—that inoculate us against the worst excesses of stress."

And so, the lonely boy, spurned by his peers for his foreign ways and disturbed by the social unrest that surrounded him, the boy who spent innumerable hours on the vortex of a train platform (people here, then gone, seen, then disappeared), saw, at the age of six, a ghost.

Steiner had his first clairvoyant experience—or, as a cynic might call it, hallucination—when a familiar-looking woman seemed to appear at the train station and speak to him, begging him to help her. He would later discover her identity: a close relation who had committed suicide on the very day he had "seen" and "heard" her.

In sub-Saharan Africa, those who claim to truly hear the voices of their ancestors are considered blessed. In the western world, these people are considered schizophrenic. Or simply lonely: according to a 2008 study published in the journal *Psychological Science*, people who lack social connections—as the lonely young Rudolf certainly did—have a stronger tendency than their peers to believe in the supernatural and metaphysical.

There must have been something in the water in late-nineteenth-century west-central Europe. Albert Einstein, born in 1879 in the German Empire state of Württemberg, was a visual thinker who saw complex calculations as images in his mind and struggled to translate them into the universal languages of words and mathematics. And Nikola Tesla, an ethnic Serb born in 1856 in the Kingdom of Hungary (now Croatia), not far from Steiner's place of birth, was another famous visuospatial thinker, who would picture his inventions in complex detail before ever working them out on paper.

Unlike Tesla, Steiner didn't have visions of alternating-current electricity or turbines or radios. Unlike Einstein, he didn't "see" a theory of relativity. But he surely had a photographic memory (he reportedly never referred to notes when delivering lectures), and his thinking was similarly eidetic. In his autobiography, he claims that he lived more in the images in his mind than in the material world that surrounded him until he reached the advanced age of thirty-five. So in addition to communing with ghosts, this vividly imaginative child developed an early love for geometry.

For this young visuospatial thinker, geometry was a place where mathematics and spirituality met, where the human mind's unique ability to conceive of and manipulate abstract forms might just be proof of man's mental access to the spiritual world.

Rudolf discovered geometry as a precocious eight-year-old through his first mentor, an assistant teacher at the one-room schoolhouse at Neudörfl (where Johann was now stationmaster). His next mentors were a priest, who introduced Rudolf to astronomy, and then a doctor, who shared a love of the great figures of German literature—most notably Schiller and Goethe—with the boy. Later, at Realschule, a technical academy, the eleven-year-old Rudolf cultivated the friendship of the school's scholarly if misguided headmaster, who rather hopelessly refuted Newtonian physics, as well as the math and physics teacher and geometric drawing instructor. (At this point, contemporary readers might be seeing a budding homosexual in Steiner, perhaps a fair assessment given that his biographers don't believe that his later marriages to women were ever consummated.)

But Steiner was an autodidact on the subject of philosophy: he pasted Immanuel Kant's *Critique of Pure Reason* into the pages of his history textbook (and still went on to ace his history exams), decided he disagreed with Kant's premises, then proceeded to immerse himself in the work of other German philosophers. In his copious free time, he tutored his schoolmates, apprenticed as a bookbinder, learned stenography, and tended a grove of fruit trees with his siblings. If he were to fill out a modern-day college application, the "extracurricular" section would be black with ink.

In 1879, the prodigy enrolled at the Vienna Institute of Technology, cross-registering at the nearby University of Vienna and embarking on a dual course of study, soaking up the sciences and humanities simultaneously. He immediately sniffed out a new mentor, the philosopher Franz Brentano, whose anti-Kantian theories appealed. Brentano postulated that because we choose to, say, create art, this type of thought can't be explained by the same mechanical processes in our brain that cause us to sneeze. Intentionality—a word that carries great weight for modern-day biodynamic farmers—Brentano argued, is what separates conscious thought from reflexive physical action. It was an idea that would stick with Steiner.

And then, during his morning commute, Rudolf befriended yet another male role model. This one, though, wasn't a scholar; he was a traditional herb gatherer named Felix Koguzki, who traveled to Vienna by rail to sell his botanicals to pharmacies and a medical school in the city. The two spent their train rides speaking of spirituality, nature, and the old-world medicine man's craft. Steiner fell hard for the atavistic approach to healing; he would remember Koguzki fondly for the rest of his life, especially when he later founded the anthroposophic medicine and biodynamic farming movements.

Despite a childhood spent with his head in the clouds, Steiner had blossomed in college into a charismatic young man with an active social life. Immersing himself in the buzzing café and salon societies of Vienna, he mingled with prominent poets, playwrights, critics, philosophers, and novelists as well as the then-fashionable cadre of astrologers and followers of the occult.

As a young graduate, Steiner reportedly rehabilitated an illiterate hydrocephalic child, who would go on to medical school (he also tutored the boys' brothers, one of whom would later become a prominent music critic).

Meanwhile, Rudolf was producing a stunning volume of work: There was the PhD from the University of Rostock in Germany with a dissertation entitled *Truth and Knowledge* in 1891 and the publication of *The Philosophy of Freedom* in 1893. (In a nutshell, this book declared that spiritual freedom can only be achieved through the act of thinking—a groundbreaking epistemological argument.) There was a brief stint as the editor of a political newspaper. There was the 1895 book on Nietzsche, which gave Steiner the opportunity to visit the famous ailing philosopher's sickbed and spar with his famously obstinate sister, Elisabeth. Later, the young scholar would edit Schopenhauer and the novelist Jean Paul.

But all of this paled in comparison with Steiner's devotion to Goethe. Although he is today remembered for his great literary works, such as *Faust* and *The Sorrows of Young Werther*, Johann Wolfgang von Goethe was in fact a poet, playwright, novelist, philosopher, military expert, and ground-breaking scientist who had begun his career as a lawyer.

Goethe's multidisciplinary mind appealed to Steiner; but his scientific writings most captivated him. Although Goethe predated the philosophical movement called phenomenology, he was an early proponent of this way of thinking, which emphasized the power of observation over the mathematical-deductive and mechanistic models of the scientific method. Like Einstein or Tesla, Goethe describes coming to his scientific discoveries merely through mulling over archetypal images in his mind; Steiner surely thought the same way. Goethe made groundbreaking observations on light-related phenomena such as refraction and chromatic aberration that would inform Steiner's later theories on color and art; Goethe's book *The Metamorphosis of Plants* would inform biodynamics.

Steiner published his tome on Goethe, *Theory of Knowledge in Light of Goethe's Worldview*, at the tender age of twenty-five, four years after he had completed the introductions and explanatory notes to the Deutsche National Literatur publication of Goethe's scientific writings. He then landed yet another gig editing the scientific writings of the famous German poet, this time for the Standard Edition (*Sophien Ausgabe,* for the Grand Duchess Sophie of Saxony) of *Goethe's Complete Works*.

While researching at the Goethe and Schiller Archives in Weimar from 1890 to 1897, the perennially poor Steiner had moved in with an older widow, Anna Eunicke, tutoring her five children in return for a room. Then, when Steiner had finished his Goethe project and moved to Berlin, Eunicke moved her household to be with him and, in 1899, marry him. It was a bond of convenience rather than romance; she was 46, he, 38. Their pact called for Steiner to play the role of the scholarly father figure in return for a warm home and meals.

In Berlin, while editing the failing weekly *Magazine for Literature*, Steiner took on side gigs lecturing to various organizations. Adopting the then-unusual practice of inviting attendees to participate in the dialogue, the Austrian quickly became one of the most sought-after speakers in Germany, packing one venue with an audience of seven thousand. By 1900, Steiner's spiritually infused speeches had attracted the attention of the Theosophical Society, an occult group something akin to modern-day Scientology in its hold on the wealthy and prominent. Thrilled at the opportunity to speak openly about his spiritual experiences, Steiner parted ways with the *Magazine for Literature* and became general secretary of the Theosophical Society for Germany, Switzerland, and Austria-Hungary in 1902.

This was the age of the occult revival, an era of tarot cards and Ouija boards in people's parlors, secret societies and Gothic-revival novels like *Dracula*. Sir Arthur Conan Doyle was supporting psychics; Carl Jung was studying alchemy and astrology; Agatha Christie's mother was a devout theosophist and dabbler in the occult. From *The Strange Case of Dr. Jekyll and Mr. Hyde* (1886) to *The Waste Land* (1922)—in which T. S. Eliot mentions "Madame Sosostris, famous clairvoyante"—Steiner's birth and death were bookends for a far-out period in popular culture.

Theosophy, translated from the Greek as "god-wisdom," wove together the spiritual teachings of Egypt, India, and the Himalayas into one proto-New Age movement that appealed to Steiner's catholic sensibility. But while the theosophists were critical of Christianity in favor of eastern mysticism, Steiner was a staunch Christian mystic, very much in the tradition of Emanuel Swedenborg.* Additionally, he was careful to distance himself from the legacy of the movement's recently deceased cofounder, Helena Petrovna Blavatsky. Her claims that she was a medium with paranormal powers—able to levitate, read minds, and channel the words and thoughts of dead souls—were, to Steiner, sensationalist and beside the point.

Nonetheless, as he rose through the ranks of the Theosophical Society, Steiner's other speaking engagements dried up. Even if everyone *was* having séance parties in the privacy of their own parlors, Steiner's open embrace of the movement was not welcome on his professional lecture circuit. Anna Eunicke, too, was unhappy with her husband's move from pure scholarship to occultism. The two separated in 1906, and Eunicke died in 1911.

That same year, in a tragicomic interlude, Steiner became involved with a fraternal order called the Ordo Templi Orientis or OTO. (Steiner was a fervid believer in gnosis,** or the notion that humans have immortal souls and can access an ageless, esoteric wisdom, and may have been beguiled by the group's ecclesiastical arm, the Ecclesia Gnostica Catholica.) Only after agreeing to head an OTO lodge did Steiner discover that the final phases of initiation

---

* Steiner was a fan of Swedenborg and lectured on the Swedish spiritual leader.
** An ironic aside: In a text dating to about 200 BCE, a southern Mesopotamian Mandean Gnostic bewails the "persecution" of being forced to plant and harvest according to the Zodiacal constellations.

into the group's highest ranks called for sex-magic rituals. It wasn't a good fit for Steiner, a "pure person" like one of his idols, Emanuel Swedenborg, who, in Steiner's words, "used ... sexual energy to see spiritual worlds." He parted ways with the order.

The next year, the colorfully named theosophical leader C. W. Leadbeater happened upon a gorgeous twelve-year-old Brahmin street urchin while traveling in India and proclaimed the child to be a deity on earth. Suspecting that Leadbeater was a pedophile, Steiner was understandably revolted. (The boy, Jiddu Krishnamurti, would grow up to disavow his own beatitude, but did become an influential speaker on subjects such as peace, meditation, and society.) Steiner's biographers surmise that Leadbeater's move was intended to distract the European theosophical membership from the increasingly anti-theosophical attitude of the charismatic, Christian-leaning Steiner. In any event, a final falling-out with theosophist leader Annie Besant followed. Steiner then took twenty-five hundred of his followers with him to form his own Anthroposophic Society in 1913.

Many of the esoteric elements of theosophy lingered in Steiner's writings. For example, he was quite earnest in his belief that he had access to the Akashic Record, a kind of timeless spiritual Internet. It was from this source that he claimed to glean intimate knowledge of, for example, the prehistoric civilization of Atlantis, a society of magical powers and vehicles that hovered in mid-air. But where theosophy was a mystic movement, based on the acceptance of miraculous visions, Steiner's anthroposophy was based on conscious thought. Its title translates as "human-wisdom," and Steiner envisioned it as a philosophy rather than a religion. He emphasized the individual's powers of cognitive contemplation, and held the natural sciences in as high regard as metaphysical matters. He also believed that actions speak louder than words, which might give us license to ignore his more off-the-wall assertions. The theosophical mystics were, he maintained, fundamentally indolent. His new anthroposophic movement would be one of action. And so he set to work.

He designed and oversaw the construction (1913-20) in the Swiss town of Dornach of the remarkable Goetheanum, an Expressionist-style timber-and-concrete edifice with a large rounded dome and Beaux-Arts details, named after Steiner's hero. As building proceeded, the First World War broke out. The Goetheanum—looking like the perfect setting for a Goethe fairy tale—became a place of refuge for people of all nationalities looking to escape the war.

In 1914, Steiner had married his assistant and avid disciple, the aristocratic Russian-born, Paris-trained actress Marie von Sivers.* By this time Steiner had embarked on an indefatigable schedule of writing, lecturing, publishing,** and planning that he would follow for the rest of his life. Collaborating with von Sivers, he invented a new form of dance-communication still practiced today, called eurythmy; the two mounted productions (starring von Sivers) of the works of their friend, the Parisian playwright Édouard Schuré, as well as Steiner's own plays.

In 1919, as Germany considered how to rebuild itself, Steiner published an influential treatise entitled *The Threefold Social Order*, which called for a just and equitable society working for the common good. A Union of Threefold Order grew out of this, attracting the ire of the Marxists as well as the proto-Nazi nationalists, who went so far as to try to assassinate Steiner in 1921. But it also attracted admirers like the novelist Hermann Hesse, who joined Steiner in his appeal for postwar rebuilding and reconciliation. Other supporters of the Union of Threefold Social Order included influential business tycoons, who in 1920 joined with Steiner to start up a corporation called Futurum. This conglomeration of anthroposophic business enterprises included factories and a bank; it aimed to show that industry run according to the principles of the Threefold Social Order could benefit the greater social good and still reap profits. It eventually failed.

In 1919, one of the aforementioned tycoons—Emil Molt, the owner of the Waldorf-Astoria cigarette company in Stuttgart—asked Steiner to create a school for the children of his employees. Within three years the population of the first Waldorf School had ballooned to eleven hundred, with prospective students being turned away. In contrast to the strict schools of the time that emphasized rote memorization, Steiner's style of education, like its contemporary the

---

* Some references spell the name "Sievers," but "Sivers" is more common.
** In 1903 he established the oddly named journal *Lucifer-Gnosis*, which would later morph into a publishing house founded by Marie von Sivers, now known as the Anthroposophic Press.

Montessori method, advocated a gentler approach that emphasized artwork and experiential learning. Steiner also established schools for the developmentally disabled at this time; his concept of residential communities for special-needs students eventually became known as the Camphill Movement.

In 1921 Steiner joined forces with Ita Wegman (a female physician in a male-dominated field) to create anthroposophic medicine, a form of holistic health care that advocates homeopathic herbal remedies and alternative healing practices—such as massage and art therapy—in combination with conventional surgery and medications.* The partners established research laboratories and clinics as well as a homeopathic pharmaceutical and cosmetics company, Weleda, based in Switzerland.

Throughout his life, Steiner had been an avid naturalist, who could identify any plant or animal by its taxonomic name and describe its habits; at one point he gave a lecture on bees that's still a classic reference for beekeepers. As he was steeping himself in the study of herbs and homeopathic therapies for his work in medicine, he was struck by what he was hearing from farmers: that the earth, too, was sick and in need of healing.

Steiner began lecturing on farming in 1922. In 1924, he delivered a series of eight lectures, entitled "Spiritual Foundations for the Renewal of Agriculture," to a group of landowners who were disturbed by the ill health effects that they were seeing in their crops and animals in the industrial age. Steiner's agriculture speeches and essays are difficult to read; they waver from the general to the specific, from the tangible to the otherworldly and weird, as in "gnomes, undines, sylphs and fire spirits are actively involved in plant growth."

They're also eerily prescient. In 1923, for example, he predicted that cattle would go mad if they were to be fed meat. The next year, pausing amid a discussion of composting, he sounded like a contemporary organic-farming advocate: "All this takes a certain amount of work, of course, but if you stop and think about it, it actually takes less work than all the fooling around in chemical

---

* It's interesting to note that, in some ways, anthroposophic medicine is more open to modern medicine than biodynamic farming is to modern agriculture.

laboratories that goes on in the name of agriculture, and which also has to be paid for somehow."

Following the lectures, the farmers formed a working group to further develop Steiner's ideas through trial and error. And so was born the first anti-industrial, anti-chemical agricultural movement, predating the modern organic movement by a couple of decades.* Demeter, the first green certification program for biodynamically grown foods, emerged just four years later.**

Just as eurythmy would not exist without Marie von Sivers, anthroposophic medicine would be nothing without Ita Wegman, and the Waldorf schools were jump-started by Emil Molt, biodynamics were not created by Steiner alone. If genius is 1 percent inspiration and 99 percent perspiration, it is biochemist, soil microbiologist, and agronomic researcher Ehrenfried Pfeiffer who should rightfully be deemed the genius behind biodynamics. It was Pfeiffer who researched and tested the preparations (which were essentially homeopathic herbal treatments for farmland); his name lives on among practitioners who use his "Pfeiffer Field Spray" and "Pfeiffer Compost Starter."

A mere twenty-five when he embarked on his life's work, Pfeiffer would go on to become the director of the largest biodynamic farm in Europe, the eight-hundred-acre Loverendale Estate at Walcheren, Holland; establish the Biodynamic Research Laboratory in 1933 in New York; and cofound the Biodynamic Farming & Gardening Association in the United States in 1938. Since Pfeiffer's death in 1961, other researchers have taken the mantle, most notably Maria Thun, who developed the widely used barrel-composting method and whose astrological planting calendar is an essential for biodynamic practitioners.

As for Steiner, the agriculture lectures capped off a dizzying couple of years of achievement. On New Year's Eve, 1922, a fire—

---

* Walter James, Lord Northbourne, the father of the organic movement, was an admirer of Steiner's and a biodynamic farming practitioner, as his son writes in *Of the Land & the Spirit: The Essential Lord Northbourne on Ecology & Religion,* ed. Christopher James and Joseph A. Fitzgerald (Bloomington, IN: World Wisdom), pp. xix-xx.
** According to Demeter USA, the Demeter International certification process commenced in 1928.

possibly electrical, as it started inside the building's walls, or possibly arson, as German nationalists had made threats against Steiner and his community—destroyed the Goetheanum. (His biographers tell us he'd harbored dark premonitions about the fate of the edifice all along.) Rudolf responded to this huge setback with renewed vigor. He immediately got to work designing a second Goetheanum, this one built of reinforced concrete (a novel concept at the time) rather than wood and concrete. It was larger and more austere than the first, strikingly expressionist in its sculptural form and artistic intent, and it remains a source of inspiration for architects today.

This streak of achievement continued throughout 1923. At one point that year, visiting a stone circle in Wales, Steiner tossed out a remark that scholars have clung to ever since: he suggested that the circle was an astronomical calendar. Oh, said the archaeologists. Thanks.

For the first nine months of 1924, Rudolf traveled and lectured ceaselessly, often speaking on multiple subjects in a single day. He counseled an endless stream of visitors when he was at home, which was rare. And although he took care of himself, restricting himself to a vegetarian diet and abstaining from smoking and drinking (always an interesting point to bring up with biodynamic vintners), he eventually succumbed to cancer. Steiner died on March 30, 1925.

The legend left behind him a profound legacy: today countless teachers, healers, horticulturists, artists, architects, and performers are influenced by Steiner's work. But some of his words leave one with an uneasy feeling in the pit of the stomach. If he claimed to eschew dogma, what are we to make of the dogmatic nature of his writings? If he claimed his otherworldly visions were open to interpretation and amendment, why did he provide such specific details?

And if he was so turned off by the theosophists' claims about miracles, why is a prominent contemporary spokeswoman for anthroposophy someone who claims to have received the stigmata?[*] Is anthroposophy a philosophy, or is it a cult? Fixate on the most outlandish aspects of anthroposophy and you'll develop an alarmingly creepy feeling about it. Ignore them and you'll anger the

---

[*] Visit www.steinerbooks.org/author.html?au=2113 to view a brief biographical blurb on the woman in question, Judith von Halle.

skeptics who are convinced that the institutions Steiner left behind are merely vehicles for evangelization. Many critics of the Waldorf pedagogy, for example, are less irked by the spiritual aspect of it than by the deceptive pretense that a very specific spirituality doesn't inform it.

The biodynamic agriculture contingent, however, has made an ecumenical and even agnostic peace with anthroposophy. Take the case of Lynne Carpenter-Boggs, the biologically intensive agriculture and organic farming coordinator for the Washington State University Center for Sustaining Agriculture and Natural Resources. Carpenter-Boggs studies biodynamic (or, as she refers to it, "biointensive") agriculture and considers herself an anthroposophist. But just as Steiner promoted a buffet-style belief system, Carpenter-Boggs takes only what is useful from anthroposophy and leaves the rest on the table. Of Steiner's daft ramblings on topics such as Atlantis, she simply says, "All spiritual leaders have spoken in parable from time to time."

For Carpenter-Boggs, anthroposophy is a method scientists can use to make the most of their minds. In the tradition of Einstein and Tesla, she thinks it's important to be able to "create a mental picture of what might be happening" in order to come up with a hypothesis to test. She suggests to her students that, if they try to think like Goethe, it might enhance their work. "I try to encourage my students to have some time with their projects where they're not trying to do A, B, or C," she says. "I tell them to 'Just sit with your plants, sit with your plots, sit with your bacteria, and watch.' It often doesn't lead to anything, but it can."

Steiner was a master of this style of deep concentration. If it unearthed dreamlike visions of other worlds, it also brought him profound insights on subjects such as architecture, philosophy, medicine, and agriculture.

"We need to really appreciate and use our minds," says Lynne Carpenter-Boggs. "There are many ways of knowing. There is logic, there is intuition, there are dreams, there is conversation, there is observation. All of these and more should be respected and developed."

CHAPTER THREE

The Banality of Cow Horns and Broomsticks

Q: Is it actually right for us as anthroposophists to
resuscitate the production of grapes for wine?
A: The question of how things *ought* to be is a difficult one
nowadays … That's why I said we should certainly take
cow horns and use them, but that to become bull-headed in
our opposition to various things could be very harmful to
the cause of anthroposophy.

—Rudolf Steiner

In August 2002, Barbara and Bill Steele purchased an overgrown
former dairy pasture and homestead in Jacksonville, Oregon. If
this were a typical vineyard story, it would proceed like this: "After
clearing the land and installing drip irrigation, they planted new
rootstock and vines in the spring of 2003." But this isn't your typical
vineyard story. This is a biodynamic vineyard story.

And it still doesn't unfold the way you might imagine.

In 2003, the Steeles began gathering temperature data after
installing weather stations throughout their 117-acre property.
Studying tables of climate conditions in various French wine regions,
they found that the *arrondissement* of Valence—the Northern Rhône
Valley region that includes the Hermitage appellation—provided a
good benchmark for their farm's microclimate.

In the meantime, they had been assembling a team of soil
scientists, architects, and contractors. In 2004, they broke ground
on infrastructure projects, moving large rocks, building roads, and
bringing in power and water lines.

Only in 2005—three years after purchasing the site—did the
Steeles begin to plant. On the strength of four decades of weather
observations made at Hermitage, they established eleven and a half
acres of syrah, grenache, viognier, marsanne, and roussanne vines in
channels of an ancient riverbed percolating with egg-shaped rocks,
similar to the famous *galets* of the Northern Rhône.

But they didn't stop there. Using as a guide a property map that's
color-coded by soil type, by thickness of soil structure, and by water-
holding capacity, they also determined which fruit and vegetable

crops they should sow in the estate's three and a half most fertile acres. Eventually, their farm will encompass fifty acres of vines and fifteen more of produce, each plant carefully chosen to thrive in the small patch of soil it inhabits.

There are two explanations for why the Steeles have been so painstaking in their approach to farming. The first is that they were determined from the beginning to farm their estate biodynamically. Which means that—with no access to crutches such as pesticides and fertilizers—there is no room for error.

The second is that Alan York is their advisor.

In the rarefied world of biodynamic consultants, York is a superstar. He had, at last count, seven clients on four continents. He travels to South Africa and South America with frequency. He advises Sting and Trudie Styler at their nearly one-thousand-acre Tuscan estate, Il Palagio. His wife looks like Goldie Hawn. York has a snow-white ponytail—the wine writer Alice Feiring refers to him as "the Silver Fox"—twinkling brown eyes, and the smile lines of someone who laughs frequently. He speaks in a pleasingly relaxed Southern drawl and curses with brio. He frequently concludes a conversation with, "That makes all the sense in the world!"

York's high-powered client list includes wine-world heavy hitters like Benziger Family Winery in California, Casa Lapostolle in Chile, and the ninety-six-thousand-acre Bodega Colomé estate in Argentina. With their paltry-by-comparison 117 acres, the Steeles don't quite fit York's typical customer profile. But they presented him with the opportunity to start from scratch, with a couple of eager students and an unplanted pasture that hadn't been stripped by chemicals. (Not that York shies away from converting conventionally farmed properties: "The whole thing is elastic to the nth degree," he says. "Look at the abuse the human body can take and still be a champion.") It was a chance for the consultant-to-the-big-time-winegrowers to show that he could make his methodical style of biodynamics work on a small scale and with a limited budget.

York chose his mom-and-pop operation wisely. The sole biodynamic wine estate in southern Oregon, Cowhorn Vineyard & Garden sits just across the road from the Applegate River in a valley surrounded by foothills of the Siskiyou Crest range in the Klamath-Siskiyou ecoregion, a biodiversity hot spot and the largest concentration of intact watersheds and roadless wild lands on the West Coast of the United States.

A dairy farm neighbors the property on the north side; Bureau of Land Management land borders its east and south sides. The wild hillsides that rise from the northern half of the farm are thick with the Douglas fir endemic to Pacific Northwest rainforests, while the southern slopes are dominated by the pines and oaks more prevalent in California.

Before they came north, the Steeles lived what Bill refers to as "a homeopathic lifestyle" in the Bay Area; they appear to have settled easily into their new bucolic life in rural Oregon. Both are tan and fit, with sparkling blue eyes and palpable energy. They lunch on hummus, apples, and carrots from their kitchen garden. Breaking with the local farming fashion—those ubiquitous Carhartts and work boots—they're dressed on the day I visit in the shorts, trail shoes, and nylon sunhats of a couple of eco-vacationers.

But they're all business when it comes to running their farm. Bill has the intuition and sound judgment of a former Wall Street analyst who got out of the game when the getting was good; Barbara commands the organizational prowess of a former CFO. "In farming, it is the responsibility of the animals and plants to bring what they have to the table and it is the responsibility of the humans to bring their brainpower to the table," she confides to me. "I know how to research the data to create an environment where we can all coexist. That's my job." With her knack for data management, look of determination, and ever-present clipboard, one can see why Barbara was asked to pour her wines recently for the likes of Warren Buffett, Carly Fiorina, and Arianna Huffington at *Fortune* magazine's Most Powerful Women Summit.

In short, this couple has a healthy respect for the power of numbers, which makes them a natural fit for Alan York. Computer printouts in hand, they nod in agreement when he says things like, "People think biodynamics is antagonistic to analytical science. Nothing could be further from the truth."

York visits the Steeles every four to six weeks during the Oregon growing season; he concentrates his time on his southern-hemisphere clients during Oregon's winter months. When I visit Cowhorn one late-May morning, I find Alan, Barbara, and Bill sitting around chatting about the weather. That is, they are sitting at the dining-room table of their home-cum-office, surrounded by stacks of thick black binders filled with temperature data, and poring over spreadsheets filled with columns of tiny numbers. As York sips a smoothie, he helps them to estimate when bud break and bloom might happen, and decide upon which day they should irrigate. On the walls are four whiteboards, inky with task lists.

"This is where biodynamics and science merge seamlessly, in good farm management," York says, turning to me. "The Old World learned over hundreds—in some case thousands—of years. In a new region, you can condense that learning curve. That's why record keeping is critical."

The discussion moves to disease monitoring. Barbara has been studying predictive models for powdery mildew, and Bill is carefully planning a defensive spray program. Even though biodynamic farms are permitted to spray small amounts of sulfur under the certification program administered by Demeter USA, Bill prefers to use stylet oil, a type of white mineral oil, instead. York agrees that it may work, and isn't too concerned anyway, since they only had to spray twice last year. "When you have that state of health, that is a successful practice," he says.

We move outside, where it's already hot, dry, and dusty by 9:30 a.m. We stop to inspect the kitchen gardens, and Barbara shows me a large glass Mason jar, its sides cracked. "We made the mistake of keeping our Preparation 500 in this jar, on the shelf of the garage," she tells me. "It must smell pretty good to plants, because a habanero pepper in that planter box there sent a root all the way through the window to get at it. It broke through the glass jar and ate up all of my preparation! And now look at how big that pepper plant is!" She laughs.

I gasp. The thing looks like the spawn of the Little Shop of Horrors. "We keep our preparations underground now," says Barbara.

York grins, dons a broad sunhat and strides purposefully through the vineyard, kneeling halfway down a row of viognier vines. The next forty-five minutes are spent reviewing the fine points of pruning and cordon training with the Steeles and their vineyard-management team.

A single cane from each vine arcs over the trellis wire; York slices off suckers and buds with a small, sharp knife, leaving an evenly spaced series of perfectly positioned buds on the cane. This will result in balanced growth: The buds will flower at the same time and produce uniformly ripe fruit. The arcing shape will expose each cluster to the maximum amount of air and light, minimizing the risk of rot and disease.

We move next to the fruit and vegetable plots, where York surveys infant cherry, apple, pear, and peach trees. And hazelnut saplings, the roots of which have been inoculated with black Périgord truffle spores. When Barbara explains that they hope to grow a crop of the precious and famously impossible-to-cultivate fungi, York's eyes twinkle and he suppresses a chuckle. The man is a realist.

Nearby, purple and green asparagus tips poke out from under brown mounds of earth; what look like scaly baseballs nestle between thistle-like artichoke fronds. The Steeles worriedly ask York what they should make of the mysterious black specks that are showing up on the spear-shaped leaves of one plant. "Caterpillar poop!" York laughs, pulling the fronds apart to reveal the furry culprit, curled up in an apparent nap. "The compost tea helped with the caterpillars and the earwigs," Barbara muses. "I guess we just didn't get this guy."

More time is spent reviewing pruning, this time amongst the stately head-trained grenache vines. Then, on our way back to the house, York takes me on a detour to see the equipment sheds and their contents: two tractors and various trailers and tillers; a plough, a weed knife. The accoutrements of biodynamic farming are here, too: a shiny steel custom-built stirring machine, an Earth Tea Brewer, and a sprayer that hooks to the back of an ATV for applying biodynamic preparations.

I'm in shock. Not at what I'm seeing, but at the state it's in.

I have visited many vineyards over the years, and I can say unequivocally that this is the … *cleanest* farming equipment I have ever laid eyes on. Do these people scrub their machinery with toothbrushes? "One misconception is that people think biodynamics are dirty," York says, wagging his finger at me with a chuckle. "On the contrary, we are anal."

It makes sense, in a way. Without recourse to chemical pesticides and fertilizers, biodynamic farmers have to be on top of their game. They have to prevent disease and malnutrition before it starts.

Which means that they must pay attention to details. Case in point: One of the underlying themes of biodynamics is the concept of self-containment, or the recycling and re-use of the site's natural assets. To achieve this requires an almost military level of organization and—yes—sanitation. Biodynamic compost can't be a moldy and stinky pile of fly-infested cow pats gone bad. It's got to be clean and pure, something you can plunge your bare hand into and take a big sniff of.

In the Steeles' garage, a giant trashcan teems with red wigglers—vermicomposters who munch through all the property's food waste, turning it into clean, humus-brown plant nourishment. All water used in the winery is tidily captured and re-used. And Barbara has taken her self-containment campaign beyond the natural resources on her property. She has convinced her local waste authority, Ashland Sanitary & Recycling, to pick up and recycle their aluminum capsules (the metal bands that wrap around the necks of wine bottles); she set up a cork recycling program at the Ashland Food Co-op. She has partnered with Wine Bottle Renew to sterilize used bottles for re-use and sends older used bottles to The Green Glass Co. to be "up-cycled" into glass tumblers. The couple donates hundreds of pounds of extra produce to local food shelters.

But Cowhorn still isn't the biodynamic ideal: for one thing, they don't keep domesticated animals, other than dogs, on their farm (there is, of course, a dairy next door), and for another, they do irrigate. Irrigation is a thorny subject: widely used to grow produce all over the world, it still isn't legal in some winegrowing appellations of Europe, where watering vines is considered to be fooling with the hand of cards dealt by God and *terroir*.

Many vineyards in Oregon do not irrigate, especially in the moist Willamette Valley and higher-altitude sites in other parts of the state. But here at Cowhorn, in the rocky, fast-draining soil beneath a hot May sun, it's obvious why the irrigation ditch surrounding the property is a godsend. Small amounts of water, slowly delivered to the vines via drip irrigation, keep plants alive during the rain-free summers. Still, Alan York grapples with this issue. "We can talk about irrigation as if it is a separate event, but the impact of irrigation affects everything," he says. "Every single thing you do is either detrimental or beneficial to managing disease and pests. It's the attention to the details that add up. A nickel here, a dime there … everything is contributing. It's really a different way of looking at management."

Later, back at the dining-room table, the Steeles tell me that they—as completely untrained in winemaking as they were in farming when they commenced this project—vinify their own wines, conferring with a consultant in California over the phone only when questions come up. "There's not one test at UC Davis that I could pass," Bill admits with a sheepish grin.

I've got a sinking feeling in the pit of my stomach as Barbara pours their latest releases. Over the years, I've tasted a lot of flawed plonk from untrained winemakers. I assume this is going to be more of the same. But a single sniff from each glass tells me that these wines are clean. A taste confirms that they're delicious. I'm incredulous.

"The grapes come into the winery in such a good state that we would have to trash them to make bad wine," Barbara explains with a shrug.

"We just don't do much to them," adds Bill. "We use native fermentation and just move them to barrel. No intervention."

York beams with pride. "The single biggest problem with wine quality is the disconnect between the vineyard and the winery," he observes.

We head back out to the vineyard and the ninety-five-degree afternoon heat, where the consultant dons his spectacles to tutor Barbara, Bill, and vineyard manager Martín Figueroa in the fine art of grafting under the merciless late-afternoon sun. As the Steeles' dogs, Buddy and Bo and Deuce, nuzzle York lovingly, he slices into the side of a rootstock, cuts a bud off a chilled grapevine cutting, and inserts the bud into the slice, closing the bark over it and tying off the top and bottom of the incision with a rubber band.

Here's how to start farming your vineyard biodynamically: Procure copies of the twenty-nine-page "Guidelines and Standards for the Farmer" and the nine-page "Demeter USA Wine Making Standards," both published by the biodynamic certification organization Demeter USA. Then purchase a copy of Rudolf Steiner's lectures on agriculture from the Biodynamic Farming and Gardening Association.

Attempt to read and understand all materials. Feel your eyes go glassy. Feel your brain go fuzzy.

There is no textbook. There most certainly are no CliffsNotes.*

At this point, you have a few choices.

You can try to go it alone.

You can enroll in classes at an educational institution such as the Rudolf Steiner College in Fair Oaks, California; the Michael Fields Agricultural Institute in East Troy, Wisconsin; or the Pfeiffer Center in Chestnut Ridge, New York. Or Emerson College in East Sussex, England; or the Biodynamic Education Centre in New South Wales, Australia, if you're willing and able to travel that far.

You can sign up for various seminars offered by the Josephine Porter Institute for Applied Bio-Dynamics in Virginia or get yourself an apprenticeship at a biodynamic farm or vineyard.

Or, if you can afford it, you can hire a biodynamic consultant.

Which is what most vineyard owners end up doing, because they're too busy growing grapes and making wine to pursue any of the other options. So they pay said consultant large sums of money (and grouse about that money) and they do what the consultant tells them to do.

What does a biodynamic farm look like? For starters, a lot like an organic farm: Free of synthetic fertilizers and pesticides. Heavy on compost. Green with cover crops. You *can* spray minute amounts of sulfur and copper to prevent fungal diseases. You *cannot* plant genetically modified organisms.

Here's what differentiates your biodynamic farm from your neighbor's organic farm: Your credo.

Robert Gross, owner of Cooper Mountain Vineyards in Beaverton, was the first Oregon vintner to embrace biodynamic agriculture when he retained Alan York in 1995. "We were transitioning to organic certification at that time and somebody came by and told us what we had to do to make the grapes organic," Gross recalls. "When Alan came, the first thing he said was, 'Bring me a shovel.' We were

---

* Jim Fullmer, executive director of Demeter USA, was compiling a workbook for biodynamic certification candidates as we went to press. The project, entitled "Biodynamic Farm Standard Workbook and Agricultural Professional Training Program," was due to be completed sometime in 2011. Until such a workbook is published, biodynamic consultants will continue to enjoy great demand for their services.

scratching our heads. What did he need a shovel for? Well, it turned out that he could tell how healthy the farm was by looking at the soil. So we went through the vineyard with a shovel, and he was counting earthworms and looking at how compacted the soil was."

In short, you're not just managing a crop. You're not just managing a farm. You're managing an ecosystem. It's a giant, farm-sized organism, of which every bird, bee, and bug is an integral part. Everything—from the stray deer to the stray dandelion seed—contributes to the dynamic circle of life. Worms are very, very important.

Here in the Pacific Northwest, we're familiar with the idea of the ecosystem. Like an old-growth rainforest sustaining thousands of species, the biodynamic farm should be self-sufficient. It should require few if any inputs from the external world.

Where the mindset of organic cultivators is "do no harm," biodynamic goes one step further. Its practitioners aim to do good. Biodynamic-certified farms should be "regenerative rather than degenerative," according to Demeter's guidelines.

At least 10 percent of every biodynamic property is reserved for natural habitat. This isn't merely for wildlife-preservation purposes; it's for self-preservation as well. Buffer zones protect farms from pests and disease by establishing borders between concentrations of susceptible crops and by creating habitat for beneficial plants, insects, and animals whose natural function is to combat agricultural invaders.

Soil nourishment comes not from organic fertilizers, but rather from carefully husbanded compost piles, consisting of plant debris, winemaking pomace, office refuse, and manure sourced from local organically raised animals (preferably animals living on the same farm).

Weed killers—even organic ones—are not permitted. Instead, biodynamic farmers study weeds and try to understand them. As Ehrenreid Pfeiffer wrote in his book *Weeds and What They Tell*, these unwelcome plants might be nature's message to us that something is out of balance in the field. Weeds can be outwitted with soil-sustaining cover crops and carefully timed tilling and mowing. And, when all else fails, the old-fashioned way: pulling them by hand. Which is when it's useful to have a few farm animals around, such as sheep that will wander down your vine rows, yanking up unwanted green matter, dropping manure as they go, and with any luck, avoiding getting tangled up in your trellis system or knocking buds off the vines.

And then there is the striking irony that many of the things we consider to be weeds, at least here in the Pacific Northwest, are gold to a biodynamic practitioner. The same noxious nettles, horsetail, and dandelions that are the bane of most farmers and gardeners are carefully cultivated by their biodynamic neighbors for their homeopathic healing properties.

Which brings us to the subject of the preparations, or what Bill Steele refers to as the "everything else" of biodynamic farming. These plant panaceas are said to deliver nutrients to vegetation and soil in a way that no fertilizer can, while also protecting against disease, increasing sun absorption, and performing countless other wonders, both physical and spiritual in nature. The preparations— or "preps"—are most often referred to by their code numbers, 500 through 508. These numerals are a convenient shorthand for those moments when the full explanation—"Could you go dig up that skull with the oak bark in it?"—sounds excessively verbose and a bit ridiculous. Numbers 500, 501, and 508 refer to field sprays; 502 through 507 are compost additives.*

Farmers are encouraged to make their own preps, but most busy winegrowers only produce a few, if any, on site, preferring to purchase them from the Josephine Porter Institute or a local supplier.** In addition, members of local biodynamic gardening associations often join forces to produce the preps in bulk batches. Given the bizarre assemblage of animal body parts required to make them, it's easy to see why prep making is one task biodynamic vinegrowers don't feel too guilty about outsourcing. To our modern-day minds, the notion of stuffing a deer bladder with yarrow and hanging the stinky thing from the rafters all summer just sounds a bit far-fetched.

---

* See Introduction for detailed descriptions of each preparation and for a footnoted explanation of the preparation numbers.
** This looks as if it breaks the "no or few inputs" rule of biodynamics. To this, Jim Fullmer, executive director of Demeter USA, responds that no biodynamic practice is perfect. And biodynamic consultant Philippe Armenier suggests that the whole nation is, in a sense, one giant farm …

And the preps are the part of biodynamics that stops most observers in their tracks. Are these farmers truly stuffing dandelions into bovine stomachs and drowning sheep skulls in rain barrels? Come on, now. Seriously?

Practitioners reply with a shrug that the preps seem to work and the animal parts just seem to make the best vessels. And although the prep-making sounds like horticultural witchcraft to most people, wine purists tend to get it: Those who appreciate the value of wooden fermentation tanks and cork-topped bottles understand the biodynamicists' need to ferment their preps in natural, semi-permeable containers. They also appreciate the observation of traditions. Biodynamic farming traces its roots back to a time when animal parts were easy to come by and nothing was wasted. Bones became flutes; horns and internal organs were the Ziploc bags and Tupperware containers of their day.

Steiner's explanation was, of course, more convoluted: "A cow has horns in order to send the formative astral-etheric forces back into its digestive system, so that much work can be accomplished there by means of these radiations from the horns and hoofs," said the Austrian sage. "Anyone trying to understand foot-and-mouth disease—that is, how the periphery of the animal works back on the digestive tract—needs to understand this relationship ... In a horn you have something that can radiate life, and even astrality ... If you could crawl inside the living body of a cow ... you would be able to smell how living astrality streams inward from the horns."

He lost me at "astral-etheric." But the artist who carved the *Venus of Laussel* twenty-five thousand years ago might well have understood what Steiner was talking about. To this Paleolithic sculptor, the resemblance between a bison horn and a crescent moon was a striking, and perhaps heaven-sent, message. Indeed, the fixation with bovines has solid historical precedent in the annals of the eastern religions Steiner so admired. In the *Gatha Ahunavaiti* verses—a sacred text supposedly composed by Zoroaster himself— the term *gēush urvâ*, meaning the soul of a cow, is used figuratively to refer to the spirit of the earth or of creation. And any visitor to India quickly comes face to face with the Hindu belief that the cow is a sacred symbol of life. In addition to allowing her to hold up rush-hour traffic, the Hindi people reverentially drink the cow's life-giving milk and refuse to slaughter her or eat her flesh.

"Why the cow was selected for apotheosis is obvious to me," wrote Mahatma Gandhi. "She was the giver of plenty ... Hers is an unbroken record of service which does not end with her death. Mother cow is as useful dead as when she is alive. We can make use of every part of her body, her flesh, her bones, her horns and her skin." Not to mention her poop. The key to biodynamic farming is compost, the bringer of life to soil. And the key to this compost is life-giving cow manure. The cow pat is a wondrous substance, brimming with microorganisms, quick to sprout seedlings, attractive to earthworms, flies, and dung beetles, and repellant to mosquitoes. Gardeners swear by it because its relatively large proportion of organic matter conditions and builds soil while slowly releasing balanced levels of the essential nutrients nitrogen, phosphoric acid and potash (aka N-P-K).

As wine writer and biodynamic vigneron Monty Waldin describes it, Steiner saw the dung-fueled compost pile to be itself a sort of living organism, "with the preparations representing the pile's own organs: yarrow as the lungs, breathing in cosmic influences; chamomile as the stomach, making sure the mix of elements within the pile and soil are digested and processed correctly; stinging nettle as the liver, its influence being to cleanse; oak bark as the brain, reining in excess; dandelion as the inner body or self of the pile; and valerian as the blood, bringing warmth, stimulating life."

In order for a farm to be Demeter certified, biodynamic compost must be applied at least once every three years. Compost piles vary in volume, depending on the size, output, and needs of the farm. Beginners can jump-start their compost programs by inoculating them with the Pfeiffer Compost Starter, a blend of all the compost preps (502-507) plus 500 and additional microorganisms. Those without sufficient space or resources to create a massive heap can make Maria Thun's Barrel Compost (also called "Biodynamic Compound Preparation"), a compact compost composed of cow manure, stinging nettles, egg shells, basalt rock dust, and all the compost preparations. After fermenting the barrel compost in a brick-lined pit for three months, the farmer dilutes and sprays this mixture directly onto the crops instead of spreading it over the soil, as he would with a traditional compost pile.

Speaking of spraying, let us turn our attentions to preparations 500 and 501, the cow-horn preps, liquids which must be sprayed over every vine in the vineyard at least once annually. With these we encounter another of the biodynamic mysteries: the dilution process. Why must these substances be stirred so vigorously, and for so long?*

To "enliven" the liquid, Steiner advises, "you have to start stirring it quickly around the edge of the bucket, on the periphery, until a crater forms that reaches nearly to the bottom, so that everything is rotating rapidly." In fluid dynamics, this vertical cone created by the circulation of fluid is called a "vortex"; Jules Verne, Edgar Allan Poe, or Herman Melville would call it a "maelstrom." It's easy to see why practitioners speak of the stirring action as one of "dynamizing": the momentum of the water, like a curling wave turned sideways, creates a tremendous amount of energy. Once this whirlpool effect is achieved, Steiner continues, "Then you reverse direction quickly, so that everything seethes and starts to swirl in the opposite direction. If you continue doing this for an hour, you will get it thoroughly mixed." Order is broken into chaos; then order is restored.

My initiation into the stirring of the preps took place in early May of 2003, when the charismatic vintner Jimi Brooks took me on a tour of the fifty-five-acre biodynamically farmed portion of Momtazi Vineyard (the remaining acres were, at that time, farmed organically). Lurching over uneven ground in his uninsulated, military-style Land Rover 90, we drove to inspect his steaming mounds of compost, his groves of nettles and horsetail, and finally, his stirring shed.

Now here's where biodynamics play tricks with a person's memory. The practice was so new to me then that I could have sworn that there was a giant black witch's cauldron in that shed. But really, I rather think in retrospect that it was a plain old barrel. His stirring implement was a handmade switch broom hanging from the ceiling. Jimi struck me as a spiritual person, but he also had a wicked sense of humor. If biodynamics sounded like witchcraft to some people, so be it: he would make like a witch and stir his preparations with a broom.**

---

* As described in the Introduction, tiny amounts of solid material are stirred ritualistically, for an hour.
** The Australian biodynamic evangelist Alex Podolinsky describes the classic central European stirring implement as a willow broom hanging from the ceiling in his treatise, *Living Knowledge* (2002). Other biodynamic

A strapping, broad-shouldered former football player, Brooks made stirring look easy when he only had fifty-five acres to worry about. Tragically, he died of an aortic aneurism at the tender age of thirty-eight in 2004. Momtazi Vineyard has changed since then; now all 240-plus acres under vine are farmed biodynamically.* Today, with Brooks no longer around to stir, and much more ground to cover, Moe Momtazi—an engineer, after all—has declined to purchase a cushy custom stirring machine like the one at Cowhorn. Instead, he has designed and erected two unusual water towers to do the job. Water pumped to the top cascades down through what appear to be a series of oddly shaped double-basin sinks. As it hits each bowl, the force of the water sends it swirling clockwise, then counterclockwise, and so on.

What is it that is so familiar about the shape of those twin oval basins, optimally shaped for creating vortices that will dynamize the water? In fact, they are the molded plastic impressions of an extremely pregnant woman's belly. (The lady in question is Sally Lammers, the wife of Jim Fullmer, executive director of the Demeter USA.)

So forget all that silliness you've heard about witch's brooms and black cauldrons. That's old folktale stuff. Nowadays, modern biodynamicists simply send their preparation water through a "flow form" consisting of—as Jim Fullmer's daughter so aptly put it— "mama bellies." What's so strange about that?

By contrast, the creation of the first ever batch of preparation 500 in 1922 was totally banal. Recalling the occasion later, Ehrenfried Pfeiffer described Steiner's instructions as casual: "Just 'do this and then that.'" When it came time to unearth the cow horns in the summer of 1923, Pfeiffer and his colleagues found that they had forgotten to make note of where they had buried them, so the somber event turned into a comedy of errors as a good deal of time was

---

texts describe the broom as a simple way to disperse the preparation: dip the broom in the liquid and shake it over the plants to scatter droplets of fluid.

* There are nearly three hundred additional acres of forest, aquifers, and animal pastures. A couple dozen sheep work the vineyards, pulling weeds out by their roots; a couple dozen more longhorn—a nod to Moe Momtazi's college days in Texas—cattle provide, when they meet their demise, super-sized cow horns.

spent digging around in search of them. Just as Steiner was about to leave for an appointment, a cow horn was found. The philosopher unceremoniously dumped its contents into a bucket of water and demonstrated with his walking stick how the water should be stirred. No incantations. No cauldron. No broomstick.

But there *is* something vaguely magical about the process, no matter how it's performed. And that is this: practitioners claim that a prep that has been stirred for an hour starts to act less and less like cold water. It feels warmer, more viscous. It changes color, taking on a creamy white hue.

I've stirred 501 myself once, one September morning at Maresh Vineyard in the Dundee Hills, under the tutelage of Scott Paul winemaker Kelley Fox, who spoke to me about Dostoevsky as I sat on a black plastic trash bag (to keep my rear end dry) and churned the water in a tamale cauldron using Fox's favorite vineyard stick, a sturdy thing with a nicely pronged top that made for easy gripping. Fox hadn't made her own 501, but had instead purchased the preparation from the Josephine Porter Institute. An avid home baker, she confidently measured a teaspoon of the white powder in her palm and tossed it into eight gallons of water. I started to stir.

I felt vaguely ridiculous, but about midway into the hour, I did notice the water changing color, taking on a milky hue. It became much easier to achieve a vortex, taking about half as many rotations to get there. The water felt strangely taffy-like as I switched directions, and seemed to expand and froth, so that, toward the end of the hour, it sloshed over the sides of the pot more and more often. Perhaps I was just growing tired and klutzy. Perhaps all that rhythmic stirring had lulled me into a semi-hypnotic state.

Or, perhaps, that white tinge was the result of millions of trapped oxygen bubbles that—as they do in a cresting wave—reflected sunlight. Where the backyard horticulturist is instructed by organic gardening magazines to tinker with aquarium pumps to ensure his compost tea is getting enough oxygen to keep the "good" bacteria alive, the biodynamic farmer can send his tea—or his wine, not to mention his preps—for a spin through the stirring machine or flow form to deliver oxygen to his precious microorganisms.

Another question that comes up with the dynamization (stirring) process: how can a couple of grams (a "pea" or a "pinhead," as

Steiner put it) of preparation material stirred into a large bucket of water, then sprayed over *acres* of vineyard possibly have any effect? We're in the realm of homeopathic quantities here, and although a few scientific studies have found that dilution might somehow enhance the efficacy of certain substances, most of modern science disputes this. One surmises that the mere act of spraying the foliage, in the same way that an orchid lover spritzes her precious plants, must have some salubrious effect on the vines, regardless of what's in the preparation.

Time to travel to firmer ground: tea. Unlike the rest of the preps, with their arcane animal body parts, endless stirring, and burial rituals, 508 is simply a giant pot of horsetail tea; teas made from chamomile and nettles are also used, although they're not official preparations. These familiar-smelling liquids are a reminder that the biodynamic preparations are just like the homeopathic health remedies taken by humans. We drink chamomile tea because it's calming; biodynamic farmers spray it over their crops to soothe them. We drink nettle tea to fight off intestinal disorders, allergies, and skin problems; plants drink it to fight off irritants. Horsetail tea cools fevers and soothes inflammatory ailments in humans; it prevents fungal diseases in plants. Order a preparation through the mail and it often arrives in the form of petite white tablets, just like the homeopathic pills you might purchase at a natural-foods store.

"It's important to note that we never get into inorganic chemistry; everything stays within the realm of the living," Steiner advised. And so the preps must be stored in glass jars or pottery crocks with loosely fitting lids to allow for air flow. Visit a biodynamic farm and the owner will proudly show you his wooden chest insulated with moist peat, filled with properly stored and clearly marked preps and dried teas.

It must be noted that the "realm of the living" doesn't necessarily mean "alive." For example, biodynamic practitioners ward off troublesome varmints such as gophers, ground squirrels, and even weeds by burning one of the unwanted pests and making a tea from its ashes or simply sprinkling the soot directly over the affected area. The general idea behind this macabre practice is that the stench of the death of one's own species—whether vegetable or animal—will keep newcomers away.

When you add up the stirring and the storing, the compost management and the prep making, the hoeing and the birdhouse making, the data collection, the record keeping, the animal torching, and the endless reams of reading, biodynamic agriculture just looks like a heck of a lot of hard work.

It's also not cheap. Although biodynamic practitioners claim that their careful farming practices result in long-term savings, the short-term costs, both in time spent and in monetary demands, are evident: one vinetender estimated to me that managing a BD vineyard is 10 to 15 percent more costly than managing a sustainably farmed property. And that's before hiring a consultant, at something like a thousand dollars per visit.*

Then there's the Demeter certification process. New applicants paid a $480 inspection and paperwork fee in 2010, and annual renewals were $380; when inspectors' travel expenses aren't covered by their stipends, there are additional costs. On top of this, renewing applicants are subject to an annual licensing fee of 0.5 percent of gross sales, or $5,000 for every $1 million earned. It's not prohibitive, but it's a hefty chunk of change for a struggling small business to invest in a certification that most American consumers have never heard of. For the same price, organic certification sounds like a safer bet.

The payoff, if you pass inspection, is permission to use the words Biodynamic® and Demeter® as well as the Demeter logo on your Web site or wine label. (And, if you've filled out the proper additional forms and fees, you can call your product "organic," as well.)

You might be wondering at this point if I've missed something. What about the movements of the planets and the phases of the moon? Isn't celestial timing central to biodynamics?

Yes, and no.

Virtually everyone who practices biodynamics times his or her agricultural activities to an astronomical timetable. Popular publications that explicate the phases of the moon and the positions of the thirteen constellations are *The Biodynamic Sowing and Planting Calendar* by Maria and Matthias Thun,** *Stella Natura: Working with*

---

* Different consultants offer different services; some spend more time with their clients than others. This number, therefore, is a generalization.

** The founder of the modern biodynamic calendar is German biodynamic

Cosmic Rhythms, and Brian Keats's *Northern Hemisphere Astro Calendar.* These calendars are filled with symbols of the zodiac, apparent proof that biodynamic practitioners are indeed participating in some New Age hippie cult. But let's not confuse horoscopic astrology—whereby the stars will tell you whether you'll meet the love of your life this week—with sidereal astrology, or the astronomy codified by Ptolemy in antiquity, when the lines between science and spirituality were blurry.

The reasons so many vinetenders choose to follow these guides are diverse. For starters, heavens-sent directives eliminate some of the hand wringing ("should I harvest today or wait until tomorrow?") that tends to accompany picking and planting decisions. Even vignerons who have no interest in biodynamics can be found racking their barrels during a "fruit" phase—when the moon moves in front of the constellations of the ram, lion, or archer—and under a descending moon, to prop up the wine's fruit flavors and keep its aromas from dissipating. They've got to rack sometime, so why not pick the date their favorite French forefathers would have chosen? But this is a classic American tradition as well: take a look at a current copy of *The Old Farmer's Almanac* and you'll find multiple references to moon phases, constellations, and astrological signs. Like a sailor who prefers celestial navigation to radar, the vigneron or farmer who cultivates or processes fruit while watching the sky is simply going about his or her work the old-fashioned way.

But sentimentality and stars aside, BD practitioners believe that there actually may be some sense to the lunar calendar. We know that the gravitational pull of the moon creates oceanic tides and affects animal behavior; farmers have long reasoned that the moon must have an effect on ground moisture and humidity as well. And so preparations meant to combat rot, such as 508, are sprayed during the waxing moon while water levels are supposedly rising.

All that said, Demeter USA doesn't require any proof of celestial-based timing for certification. (The subject only merits a lukewarm mention in the organization's paperwork: "Observation of the Biodynamic calendar is encouraged as a guide for harvest dates.") Because it's not really a quantifiable farming activity, and because

---

agricultural researcher Maria Thun, who has been studying the cycles of the moon and sun and the movement of the planets since the 1950s.

the various celestial cultivation calendars differ slightly according to the interpretations of the authors, it's just not something that the organization wants to track and enforce.

The perception and the reality of biodynamics are two very different things. The lunar and astrological calendars are important, but they're not *that* important. Steiner didn't drink, but he didn't discourage winegrowers from practicing biodynamics. He wrote about mixing the preparations as though it should be an ecstatic spiritual event, but when the time came, he simply plunged his walking stick into a bucket of water and then toddled off to an appointment.

By the end of my day at Cowhorn, I have learned a great deal about pruning, hedging, tilling, grafting, and spraying. I have seen a couple of tractors and tillers that would pass military inspection. I have learned that the Steeles have just packed their compost tea into a cooler and sent it to BBC Laboratories in Tempe, Arizona, to be tested and assessed for its Species Richness Diversity Index.

I have heard virtually nothing about Rudolf Steiner. There has been no mention of astral bodies, the cosmos, or the spirit world. And despite the name of the estate, I have not heard any talk of cow horns.

There was half a moment when I looked up and noticed a few giant wood crucifixes, each as tall as a tree, scattered around the property. Surely these must be a sign of the Steeles' religious zealotry? But no. Turns out they're simply T-shaped perches for birds of prey: a screech owl, Cooper's and red-tailed hawks, and the resident golden eagle family, all of which alight on these strategically located stopping places before diving down to dine on the ground squirrels, voles, and mice that are menacing the crops below.

I ask the Steeles what, in their opinion, biodynamic farming is all about. "I think it's 60 percent canopy management, 30 percent tillage, and 10 percent everything else," Bill replies after a moment of contemplation. Barbara and Alan nod in agreement.

As I leave, Alan York is reminding the Steeles to check their grafted vines in two weeks, to make sure they have calloused. I spend the five-hour drive home musing about the banality of biodynamics.

CHAPTER FOUR

Science ... or Sci-Fi?

Neils Bohr ... was said to have had a horseshoe hanging
over his office door for good luck. When asked how a
physicist could believe in such things, he said, "I am told it
works even if you don't believe in it."
—Leonard Mlodinow, The Wall Street Journal

In autumn, when the grapes are sweet, fragrant, and at the picking
peak of ripeness, the vinetender's greatest natural enemy is a
backyard bird weighing less than three ounces. It might be tiny, but
the American robin is voracious, numerous, and nearly impossible
to deter. To ward off this red-breasted foe, vineyard managers
employ a wide arsenal of defenses: netting, reflective tape, shotguns,
stinky substances, recordings of avian distress calls, owl-shaped
balloons, hawk-shaped kites, noise-making cannons (to the chagrin
of neighbors), home remedies such as Irish Spring soap, and even
wired perches that shock the unsuspecting avians with a thousand
volts to the feet.

Kevin Chambers knows how to battle birds. For ten years, he and
his wife, Carla, owned Oregon Vineyard Supply, the largest supplier
of equipment and materials to Willamette Valley vineyard managers,
as well as Results Partners LLC, a vinetending firm that, at last count,
managed 938 acres of vineyards for fifty-four clients throughout the
Willamette Valley. Although he sold ownership to his employees in
2008, Chambers remains as CEO of both businesses.

It's safe to say that, over the past couple of decades that he has
been in this business, Kevin has tried every bird deterrent on the
market, whether selling, using it on a client's vineyard, or testing
it on his own property, Resonance Vineyard. Resonance is just off
a country road outside Carlton, up a neglected potholed lane lined
with poplar trees, past a deer fence, and at the end of a dirt drive.
At the top of the gravel road is a simple manufactured home; below
it, the steep, south-facing hillside is quilted with vines, each plant
laden with fruit and gnarly at its thick woody base. The property's
patchwork includes trees and brush as well as four ponds teeming
with frogs and salamanders. At the bottom of the hill is one of the

few oak savannahs left in the Willamette Valley, a remnant of what this area looked like before agriculture found it.

Thanks to all of the elements on this piece of land that are *not* vine, Resonance Vineyard is not troubled by pests so much as its neighbors are. The trees and thorny shrubs attract azure-backed western scrub-jays, motley brown white-crowned sparrows, goldfinches, and lazuli buntings, with their copper breasts and bright-blue heads. All sing sweetly, but they also eat mites, the bane of vineyard managers. And these winged warblers lure hawks: red-tailed, rough-legged, sharp-shinned, and sparrowhawks, which circle the vineyard screaming out their wild cries and occasionally diving down to dine on a rodent or a songbird. Most importantly, these raptors attack—and scare off—cedar waxwings, starlings, and robins, all of which eat fruit, in particular ripe grapes.

It's easy to see the circling hawks at Resonance and understand how at least one aspect of biodynamic agriculture works. It's common knowledge in sustainability circles that setting aside natural areas of native habitat on a piece of property will attract predators, and those predators—in addition to a good electric bird alarm—will take care of pests (both avian and mammalian). This isn't rocket science. It's just good common sense.

But just as I'm beginning to get comfortable with this notion, Kevin Chambers leads me past his rambling vegetable garden and points proudly at a white PVC pipe sticking eight feet straight up out of the ground. A visitor casually glancing around might not even notice it, or figure it has something to do with irrigation, never mind that Chambers doesn't irrigate his vines.

But no. Inside that PVC pipe is a coiled copper wire hemmed on the top and bottom by two copper plates. It is, Chambers explains, a "radionic field broadcaster," or a "cosmic pipe."

Say what?

And then he's off, talking about quantum mechanics and string theory. And rocket science starts to sound fairly simple and straightforward in comparison with what is going on in this biodynamically farmed vineyard with its chirping birds, circling hawks, heavy grapes, and white PVC pipe pointing heavenward.

At first glance, Kevin Chambers is all business: no-nonsense rectangular glasses with the photochromic lenses that are favored by farmers; crisp black button-down with *OVS* and *Total Solutions for Specialty Ag* stitched over the left-front pocket; jeans. But walking around Resonance Vineyard with him and his entourage—three outrageously large dogs and two cats—you do get the sense that you're in the company of a zealot.

"I spend a lot of time reading scientific texts," says Chambers. "I probably have twelve or thirteen books on my nightstand right now in various stages of completion. At least half of them are some study of physics; that's really what I enjoy more than anything. Because it's in the arena of quantum physics and string theory where frankly we throw out much of what we think we know about the universe. What we find as we start breaking matter down into smaller and smaller parts is that those parts no longer behave like matter. They behave like energy forms. So there is a theory in quantum physics that says there is no matter. Matter is just coalesced energy. And the way you influence matter is with patterned energy."

Ask Chambers why he works his vineyards according to the cycle of the moon, and he cheerfully poses his own question: If all living organisms consist mostly of water, is it *really* that big a stretch to suggest that the gravitational attraction that's responsible for the tides that move all the oceans on the earth twice a day might have *some* influence on plant and animal behavior?

Actually, it is. The tidal forces exerted by the Moon upon a one-meter-tall vine or wine barrel are actually around sixty thousand times weaker than the tidal forces emanating from a 175-pound man standing one meter away." Oregon State University physics professor Raghuveer Parthasarathy concurs, adding, "It's utterly implausible that there are gravitational effects of the moon on plants."

Ask Chambers whether the preparations really work and he offers this anecdotal evidence: "There have been two occasions now where I have sprayed 501* on a nice day late in September, and within twenty-four hours, the fruit started to dehydrate on vines that were looking fine before," he responds excitedly. "It was almost shocking how rapidly and dramatically the dehydration occurred." If you've

---

* For a detailed description of the preparations, including the purported effects, please refer to the Introduction.

ever tried to eat a ripe tomato that has just suffered through a rainstorm, you can appreciate why a vintner might be excited about the possibility that a certain spray could dehydrate his fruit on a soggy September day during a wet harvest season. At the very least, it could be insurance against splitting and rotting; at the very best, it might concentrate and enhance flavors.

Chambers tells me that he believes that preparation 501 works the way it does because quartz, the crystalline form of silicon dioxide (also called "silica"), increases light absorption. The silica, in effect, acts like a magnifying glass, brightening the effect of the sun on the dreariest winter days, similar to the way a silicon crystalline wafer in a photovoltaic panel conducts and absorbs light for conversion into solar-powered energy.

"What's really interesting is that conventional agriculture doesn't even acknowledge that silica is an important plant nutrient," says Chambers. "But if I were to ash that plant down there, you would find a lot of silica in that plant. Well, why is that? Science basically says there are seventeen essential plant nutrients. But if I ash that plant down there, I bet I'll get fifty to sixty elements in it. And so I think that the problem with trying to prove biodynamics through the scientific process is the scientific process. The scientific process is limited; it really is the exercise of controlling all but one variable and measuring the changes. Well, I believe that is not how nature exists. It exists as a complex interaction of dozens of variables at any one period of time. And so for me to try to isolate that one variable and study it is so artificial as to impose constraints upon the entire process that make it nonviable. That's why science can't prove biodynamics one way or the other."

However, according to Lynne Carpenter-Boggs, a Washington State University researcher in sustainable agriculture, scientists *have* researched silica's effect on horticulture. "There have been studies that have found that silica, along with other materials, can stimulate the immune response," she says. "It creates a systemic resistance, so it is similar to giving the plant a vaccination. The plant is essentially irritated and responds by producing defense compounds."

There are so few farmers practicing biodynamics, particularly in the United States, that, admittedly, the subject has not been studied rigorously. One soil scientist I spoke with estimated that biodynamic farms account for just 0.1 percent of all farms in this country. There just isn't any funding, he went on to explain, for research into such an obscure arena of agriculture. Or if there is funding, there isn't enough momentum. In 2009, the Northwest Center for Small Fruits Research named "Development and evaluation of integrated/sustainable production systems including but not limited to organic, biodynamic, and conventional" as a funding priority. The Center has yet to receive a proposal.

It can be frustrating to sort through the research that has been done on biodynamics because these studies often don't compare biodynamic with organic. A simple comparison of biodynamic and conventional farming is going to give you similar results to those of a study comparing organic with conventional farming. And what, exactly, would a researcher of biodynamic agriculture observe? As the English wine writer Jamie Goode points out, "We can see why it is difficult to discuss the theoretical basis of Biodynamics scientifically. The undefined 'life forces' aren't specified in a concrete way, and we have no means of measuring them."

Unless you focus on the preparations. These mysterious homeopathic remedies for plants are unique to biodynamics. Jennifer Reeve, assistant professor of organic and sustainable agriculture at Utah State University, is one of a handful of American scientists who have been studying the preparations and field sprays to validate their effects. In one experiment, Reeve and her co-authors, who include Lynne Carpenter-Boggs, compared lime—the conventional method of raising the pH of acidic soil—with the Pfeiffer field spray* (application rate: two ounces per acre) in a farm pasture on Lopez Island, Washington. The lime was most effective at lowering the acidity of the soil, but it also reduced the crude-protein content of the grasses. By contrast, the biodynamic field spray lowered the soil's acidity somewhat while also *raising* the crude-protein content of the forage slightly above the naturally occurring levels.

In another experiment, Reeve, Carpenter-Boggs, and their colleagues—including their mentor, John Reganold, the Regents

---

* A microbial inoculant sprayed prior to plowdown, the spray contains microorganisms that are said to help speed decomposition of organic matter and stimulate humus formation, enhancing soil structure. It was developed by Ehrenfried Pfeiffer.

Professor of Soil Science at Washington State University—composted grape pomace (skins, seeds, and stems) with straw and cattle manure. They found that the compost heaps that had been treated with biodynamic preparations showed higher microbial activity than the untreated. (Microbes essentially gather and process nutrients, releasing them into the soil; earth with a high microbe count is soft and rich with humus.)

However, these findings are balanced by a number of studies that have found that biodynamic agriculture has *no* measurable effect on soil health. For example, a veritable Who's-Who of biodynamic research—Reeve, Carpenter-Boggs, and Reganold, as well as the consultant Alan York and two other researchers—collaborated on a long-term study initiated in 1996 and concluded in 2003 on a merlot vineyard near Ukiah, California. As the abstract states, "[T]he study consisted of two treatments, biodynamic and organic (the control), each replicated four times in a randomized, complete block design. All management practices were the same in all plots, except for the addition of the preparations to the biodynamic treatment. No differences were found in soil quality in the first six years. Nutrient analyses of leaf tissue, clusters per vine, yield per vine, cluster weight, and berry weight showed no differences."

The team did, however, measure higher Brix levels, phenols, and total anthocyanins (that is, sugars, flavor compounds, and pigments) in the biodynamically farmed grapes in one year, 2003. They also found that, overall, the control vines were slightly overcropped in comparison with the better-balanced biodynamic vines throughout the study. Generally speaking, that means that the biodynamic vines produced less, but more flavorful, fruit. This is an ideal situation for a winegrower, who typically must "drop fruit," or prune excess grape bunches, to ensure that the grapes achieve full ripeness. It isn't as encouraging for any other type of farmer, who may simply be looking to grow as much of his or her crop as possible. Even so, the authors are circumspect, writing, "[T]here is little evidence the biodynamic preparations contribute to grape quality. The differences observed were small and of doubtful practical significance."

"Whether or not the growers believe the preparations work, they are out there putting them on and interacting with their plants and their

soil more. They are taking better care of the biological and physical properties of the soil," says John Reganold of Washington State University.

Indeed, watching a biodynamic practitioner carefully apply preparation 501 to her vines one hot morning using a hand-held spray device, I couldn't shake the image of a glamorous woman spritzing her face with Evian while flying first-class. The leaves really did seem to perk up and preen as they received their refreshing mist. Which brings up a word that recurs frequently in discussions of biodynamics: "intention." To scientists like Reganold, Lynne Carpenter-Boggs, and Jennifer Reeve, "intention" means careful stewardship of one's soil and environment, and close attention to the health of one's farm.

But for committed BD practitioners, intention is more metaphysical. "Normally on a spray sheet, you have to record everything that you spray for the government. The spray form says how many gallons per acre and what you are putting on—sulfur or whatever," explains Sam Tannahill, the biodynamic-practicing director of viticulture and winemaking for the Oregon wineries A to Z, Rex Hill, and Francis-Tannahill. "Our biodynamic spray form says not only what you put on, but also when you put it on, what the lunar phase was, what the time of day was, what were your thoughts and feelings as you were stirring, what were your thoughts and feelings as you were spraying, and what did you notice?"

Thoughts and feelings. These are not tools that you'd typically find stored in a farmer's equipment shed. But to many biodynamic practitioners, they are just as important. The belief is that the preparations aren't merely herbal treatments for plants; they're carriers of the farmers' intentions, which have been swirled into them through the powerful act of stirring. While it isn't a requirement for Demeter certification, intention is that little bit of witchcraft that separates the most committed practitioners from the unbelievers.

Intention, in this sense of the word, is not a quantifiable thing that can be measured by university researchers. But Kevin Chambers, being a scientifically curious person, has tried to find an explanation for how it might work.

He describes a moment of "epiphany" in December of 2002 when he read a book called *The Field* (no, it has nothing to do with farmers' fields; the subtitle of the book is *The Quest for the Secret Force of the Universe*), which has become a go-to book for BD practitioners trying to wrap their minds around the notion of intention. In *The Field*, journalist Lynne McTaggart pulls together an eclectic group of studies—drawn from the realm of what she calls "frontier science"—to create her own theory of everything. "It doesn't talk about agriculture. There are certainly no discussions of biodynamics," admits Chambers. "But what McTaggart does is she takes all the cutting-edge research in about eight or nine different scientific disciplines—including mathematics, psychology, and physics—and weaves this unified theory of existence. I set that book down and went, 'Oh my god, that's how it could work!'"

To prove the power of intention, McTaggart jumps from psychology—citing the observer effect, whereby the administrator of an exam can inadvertently change the behavior of the person being examined—to particle physics, in which the observer changes the path of an electron merely by trying to view it. From there she gets into quantum mechanics, in which, similarly, the observer determines the state of something simply by trying to measure it. (Hence the famous—and macabre—parable of Schrödinger's cat, who is said to be alive and dead simultaneously, until an observer opens the box and sees it to be either one or the other.)

Some three decades ago, the German physicist Fritz-Albert Popp discovered that all living things, including humans, are a source of "biophoton emissions"—that is, tiny light waves. We know that plants photosynthesize light and humans absorb the Vitamin D in sunlight, but Popp's research has shown that living things respond to and emit electromagnetic waves at the molecular level. Could these waves of energy be harnessed and directed? McTaggart proposes that they can.

In 1994, a researcher in Switzerland took advantage of the baby chicken's natural tendency to adopt the first thing it sees after hatching as its "mother" by exposing a group of chicks to a small robot. The robot was programmed to wander aimlessly throughout a room, its movements random. But when the baby chicks were placed on one side of the room in a cage, the robot hovered near them. Somehow, McTaggart hypothesizes, these tiny birds were moving a machine simply by wishing their "mother" were closer.

Now consider this: If there are, as scientists like to say, as many atoms in each cell as there are stars in the sky, and if every atom is composed of charged electrons and protons, and each human body is made up of about a hundred trillion cells, we might as well just be giant blobs of energy. There is energy everywhere in the universe, even in empty space, where it is known as "zero point energy." Lynne McTaggart describes this energy as existing everywhere, all around us, in a "Zero Point Field," something like "The Force" in *Star Wars*: a vast store of ever-present subatomic energy that we energy-rich humans should be able to tap into and direct using our thoughts. Perhaps, she suggests, we already have: witness the witch doctors and practitioners of transcendental meditation who have, in rare cases, succeeded in helping others seemingly using the powers of concentration alone.

The Field is a masterfully built case for the miraculous power of intention, but take a closer look at the endnotes full of apparently reputable scientific journals and academic institutions and you'll find bastions of pseudoscience such as the Institute of Noetic Sciences, which claims to have conducted experiments that show that outcomes can be determined by intention—meditation or prayer—directed toward the desired outcome.

It's a beguiling argument, and one that faith healers have been making for generations. The popular alternative-medicine personality Deepak Chopra put it forth in his book *Quantum Healing: Exploring the Frontiers of Mind/Body Medicine* in 1989, and Rhonda Byrne called upon the power of intention in her mega-bestseller *The Secret*.

"Quantum mechanics. What a repository, a dump, of human aspiration it was, the borderland where mathematical rigor defeated common sense, and reason and fantasy irrationally merged," observes novelist Ian McEwan in *Solar*. "Here the mystically inclined could find whatever they required and claim science as their proof." The skeptic Robert Park has deemed this kind of thinking "Voodoo Science": "Quantum theory is invoked by Chopra to convey the impression that ayurvedic medicine has somehow been validated by modern science. We cannot help but notice, however, that the author of *Ageless Body* shows unmistakable signs of growing old right along with the rest of us ... Where once the magician in his robes would have called forth the spirits, the pseudoscientist invokes quantum mechanics, relativity, and chaos."

The scientist Robert Lanza concurs. "Quantum theory has unfortunately become a catch-all phrase for trying to prove various kinds of New Age nonsense," he writes. "Quantum theory deals with probabilities, and the likely places particles may appear, and the likely actions they will take." That's all there is to it. Which explains why intention really doesn't work. Because, if it did, why wouldn't every kid who has ever wished upon a star have a shiny new bicycle to show for it?

When we're under-informed about a new branch of science, we ascribe miracles to it: As I write this, scientists at Fermilab and CERN are racing against one another to isolate the Higgs boson. Popular culture has dubbed this proposed elementary particle "The God Particle." Something tells me that if it does exist and we do find it, it won't seem so God-like a century from now.

At some point in time, we will know more about quantum mechanics than the little we know now. Present theory will become proven fact. And much of the pseudoscience that has accompanied this mysterious arm of physics will, most likely, wither away. In the meantime, it's glaringly apparent that none of the research cited in *The Field* directly relates to biodynamic farming. It's difficult to see how a bunch of chicks in a room with a robot have anything in common with a vigneron and his vines.

And there's an intriguing fallacy going on here. On the one hand, biodynamic practitioners pride themselves on turning their backs on the past century of agricultural research in favor of old-fashioned traditions. They tell us that modern science can't calibrate their style of farming. At the same time, they draw from one of the most youthful and arcane branches of science, quantum mechanics, to claim that praying for their plants is a valid way to go about running a farm.

Lynne McTaggart's *The Field* isn't the only text cited by biodynamicists; it's the general notion of quantum mechanics that seems to excite everyone. "The world of wine exists in non-Euclidean space, and certainly partakes of the quantum universe; there are great discontinuities in what we know or imagine we know," writes Randall Grahm, the famous founder of Bonny Doon Vineyard in California. The French BD evangelist Nicolas Joly opens his mystifying book *BioDynamic Wine, Demystified* with a (misattributed) quote from physicist Max Planck, the father of quantum mechanics: "All matter originates, and exists, solely by virtue of a force which induces particles to vibrate."* And the biodynamic farming guru and author Hugh Lovel lectures on "quantum agriculture."

Rudolf Steiner, a man who had, as biographer Gary Lachman puts it, "a remarkable intuition about the direction that science would take in the twentieth century, with its increasing fascination with ... subatomic particles ... atomic energy and black holes," surely would have approved. The term "quantum mechanics" didn't appear until 1924, the year prior to Steiner's death, so we can only conjecture that Steiner the "spiritual scientist" would delight in the modern-day movement known as quantum mysticism. Wouldn't the quantum-based notion that we are all just amalgamations of coalesced energy floating around in a Zero Point Field fit neatly with his concept of the cosmos as a source of constant creative activity?

Steiner, who spoke so often of the opposing forces of Lucifer and Ahriman, was clearly conflicted. He rejected Kant's assertion that the human mind cannot conceive of possibilities that are outside the human experience, preferring to believe that the world is governed by unknowable cosmic rhythms and forces. At the same time, mathematics and science were fundamental to his life's work. Biodynamic agriculture is a textbook example of this conflict: on the one hand, it's practical horticulture. On the other, it's faith-based farming.

It's understandable why the most committed biodynamic practitioners have seized upon quantum mysticism to justify their more colorful practices. In essence, these farmers are practicing a kind of religion, as the title of Steiner's lectures—*Spiritual Foundations for the Renewal of Agriculture*—implies. For those who are uncomfortable with blind faith, science offers a rationalization for irrational actions. But for observers, this association with pseudoscience can be troubling. It brings to mind the current anti-science craze among the ill informed, from the vaccine avoiders to the creationists. The green-

---

* Joly attributes the statement to Planck's "Nobel Prize acceptance speech" in 1920, but it appears to be excerpted from a lecture Planck delivered in Florence, Italy, in 1944.

minded BD practitioners would be horrified to be lumped into the same group as the global-warming deniers, but the fact is that these two groups share some similarities.

In his book *Voodoo Histories: The Role of the Conspiracy Theory in Shaping Modern History* (yes, "voodoo" is a popular catch-all term for the fringe these days) journalist David Aaronovitch identifies characteristics shared by successful—that is, successfully propagated, if inaccurate—conspiracy theories. Aaronovitch cites historical precedent ("people farmed this way for thousands of years"); a group of interested skeptics ("we're turning our backs on conventional agriculture"); the tendency to "just ask questions" ("if the moon affects the tides, why shouldn't it affect plants?"); expert witnesses ("I applied the prep and saw this result; therefore the prep caused it"); and academic credibility ("Lynne McTaggart quotes a bunch of scientists who say it is possible to determine outcomes using the power of intention"). By Aaronovitch's criteria, biodynamic agriculture is just one more conspiracy.

And so we return to that white PVC pipe, otherwise known as a radionic field broadcaster. There are two joints on the pipe, each with a well in it. "You're going to have to suspend your disbelief," Kevin Chambers warns me. And then he flips up the lid over one of these joints and pulls out an ordinary zip-close plastic bag. In this bag are glass vials containing tiny white pills, resembling the homeopathic teething tablets you'd place under a baby's tongue.

Each vial is marked with its contents, measured in homeopathic "C" potencies that indicate the dilution factor. For example, the vial marked "BD 502: 12C; BD 503: 15C; BD 505: 12C" contains tablets of preparations that are mind-bogglingly diluted to concentrations as small as 10^{24} and smaller. Not every vial contains biodynamic preparations: One contains "Indigo" and "Angstrom"—diluted to 10^{20}.

"Everything has a migratory signature. Everything and everybody operates on a frequency. And what these substances do is impart a particular frequency of energy to the farming environment. These substances have an energetic signature that is picked up by the coil and broadcast over the property," Chambers explains. The copper coil in the PVC pipe is supposed to broadcast the energetic essence of these tablets—which are, as you've probably guessed, concentrated biodynamic preparations—all over the vineyard. The idea is that this energy will influence plant behavior, in the same way sunlight or water or fertilizer might. (Energy, of course, travels right through glass vials and plastic bags, which are there to keep the preparations from getting wet. Chambers even has included a map of his farm and a "biodynamic prayer"—a distillation of his intentions for the farm in a written statement—in the zip-close bag. I guess these things are supposed to be energy, too.)

"I've done a lot of things in life that people thought were crazy," Chambers once told me. "I've gone from being a lunatic to a prophet. I just had to live long enough."

The next time I see Kevin Chambers, he has dropped his Demeter credentials after six years of certification. He's still applying the preps, as he has been ever since 1997. But he can't turn his back on the siren call of science. In 2008, he cofounded a company called Willamette Cross Flow, which brings a high-tech, mobile filtration system to Willamette Valley wineries. He is also, through his work for Oregon Vineyard Supply and Results Partners, a licensed chemical and fertilizer dealer, and he has been experimenting with "high-efficiency fertilization," or the blending of traditional soluble fertilizers with organic acids, on his property. It's an alter ego that doesn't fit with the biodynamic ethos. And so the organic and biodynamic certification have fallen by the wayside: "It seems to me that the organic movement has concluded that anything made by man is unclean," he reflects in an e-mail. "I see that as an insult to our species."

No one argues that the "bio" components of biodynamic agriculture—the buffer zones, cover crops, compost piles, integration of livestock, and tillage—are not good, sound farming practices. It's the "dynamic" part that loses people. When Steiner invokes cosmic forces, or when today's biodynamic farmers refer to quantum mysticism, the conversation shifts from the methodological to the epistemological and ontological. The subject matter is no longer soil and bugs and birds. It's spirituality, intention, and energy.

It's impossible to know whether the radionic field broadcaster at Resonance Vineyard works. But there is no denying the value in the microbes percolating in the ground underfoot and the hawks circling overhead. There is some real research being conducted on biodynamic agriculture, and it's important to separate this from the random bits of pseudoscience culled from other areas of, um, academia.

"At this early stage, when there are so few studies that have been done, it may be quite noteworthy the number of positive results we have observed," says Jennifer Reeve, the professor of agriculture at Utah State University. "We have seen small differences and the question is, 'Is this a statistical fluke or is something really going on?'"

CHAPTER FIVE

The Neo-Naturalists

"I've found my most enthusiastic clients to be those chefs
and sommeliers who find in biodynamic wines a ready
partner for what they are doing in their restaurants. The
wines can be a bit eccentric, often atypical, but always
distinct and true to themselves."
> —Aaron Danforth, owner, Bon Vivant Merchants,
> Portland, OR; importer and wholesaler
> of natural and biodynamic wines

*Approximately sixty million** people log onto Facebook each day for
a specific purpose. They're not there to see pictures of their friends.
They're not there to catch up with old roommates. They're there to
farm.

FarmVille is a video game that appeals to non-gamers. The two-
dimensional graphics could have been designed by Fisher-Price. But
the look is not the point. The point is to virtually grow crops, then
sell them. Then buy more land and more animals. And plant more
crops.

The distance between urban and agrarian society is wider today
than ever before. And yet, the fascination with farming is more
passionate than it ever has been. Even as genetically modified
produce has become an everyday sight at the supermarket, so
have organic fruits and vegetables, with the implementation of the
National Organic Program in 2002.

The origins and moral implications of our food have become a
twenty-first-century cultural meme. The last decade brought us
films like *Fast Food Nation, Food, Inc., King Corn,* and *Super Size Me*
and national awareness of the raging obesity epidemic. Entertainers
like Moby, Alicia Silverstone, and Natalie Portman have become
vegan activists, while the leading lights of the culinary world
have embraced traditional butchery and whole-animal cooking.
Bookstores have stocked up on Michael Pollan's paeans to simpler
eating and Wendell Berry's classic depictions of traditional American

---

* As of January 2011.

farms. "In no other time would a highly regarded young novelist like Jonathan Safran Foer view a book about the anti-animal-eating movement as a necessary extension of his oeuvre, the way a novelist in the '60s might have felt obliged to write a book about the antiwar movement," observes *The New Yorker*'s Adam Gopnik.

Whether we're responding to the Slow Food movement's quest to connect our eating habits with the natural rhythms of our ecoregions or the environmental movement's goal of establishing a more sustainable farming system, we've gone loco for locavorism. We're frequenting farmers markets and scanning restaurant menus for the names of our favorite farmsteads. Our CSA (community-supported agriculture) boxes brim with dirt-smudged produce; vegetables and herbs sprout up in our back yards and back-deck planters. The hottest gardening trends of the new millennium have been heirloom seeds, indigenous species, and nature-mimicking permaculture. Masanobu Fukuoka's minimalist classic, *The One-Straw Revolution*, which reads strikingly like a biodynamic farming treatise, was reprinted in English in 2009,* for a new generation of ultra-farming fanatics to discover.

In sipping circles, this meme has expressed itself in the "natural wine" movement, whose spokespeople—a loose group of critics, merchants, and producers who jokingly call themselves *terroiristes*** —espouse chemical-free farming and technology-free winemaking, despite the protests of critics such as the French wine writer Michel Bettane, who claims that *"le vin bio n'existe pas!"* Filmmaker Jonathan Nossiter brought this issue to the attention of the literati with his documentary, *Mondovino*, a crowd favorite at the 2004 Cannes film festival. In Michael Moore style, Nossiter portrays über-critic Robert Parker and über-consultant Michel Rolland as villains whose evil plan is to inhabit the globe with overpriced, overmanipulated, over-flavored wines. In France, 2009 marked the first edition of the now-annual *Carnet de Vigne Omnivore: Les 200 Vins Nature*, a guide to the

---

* Fukuoka's apprentice Larry Korn, who originally assisted in translating the text from Japanese to English and toured to promote the book's re-publication, is a permaculture expert based—where else?—in Oregon.
** *Terroiristes* can be found lurking in establishments such as the two Terroir Wine Bars, New York (www.wineisterroir.com); and Terroir Natural Wine Merchant & Bar, San Francisco (www.terroirsf.com).

two hundred top natural wine producers in the nation. To be eligible, the wines must be *100% raisin*, or contain nothing but grapes.

"The natural-wine movement has been sweeping France for a few years now, with stylishly dressed millennials in trendy wine bars in affluent urban neighborhoods celebrating the peasant vignerons who defy globalization in defense of *terroir*," writes Dave McIntyre in *The Washington Post*. And while the *terroiristes* abhor spinning cones, oak chips, enzymes, and cultured yeasts, their passion is not limited to the confines of the winery walls: "Simply put, natural wine is an extension of organic and biodynamic viticulture, two approaches to winemaking that focus on the vineyard," McIntyre continues.

If there is a home for natural wine production in the United States, it has to be Oregon, where the vineyards have always been small and the yeast is typically native.* The most rustic, Burgundian, and "natural" of the Willamette Valley producers is—arguably— The Eyrie Vineyards, founded in 1966. Like any good French son, winemaker Jason Lett continues to vinify his wines naturally, just as his father David did, from organically farmed vines.**

In addition, The Eyrie Vineyards dry farms its grapes. It's a member of the Deep Roots Coalition, a collective of local vintners who don't irrigate and whose nickname (DRC) is a nod to the initials of the best of Burgundy, Domaine de la Romanée-Conti.*** Dry farming, argue the members of the Coalition, sends taproots deeper into the ground, below the topsoil, to the mineral-laden subsoil.**** It allows

---

* Occuring naturally on grapes, "native," "wild," or "indigenous" yeasts send the fruit into "spontaneous fermentation," which proponents of natural wines claim makes for a more complex wine. However, reliance on such yeasts entails risks, so commercial wineries tend to rely on purchased "cultured" yeasts in order to have more control over the fermentation process. Boosters of natural wines prefer their yeasts wild, of course.
** It's worth noting here that the Letts never bothered to pursue organic certification, and that none of the other notable Oregon pinot pioneers— among them Adelsheim Vineyard, Elk Cove Vineyards, and Ponzi Vineyards—have made any moves toward biodynamic viticulture.
*** As of April, 2010, the Deep Roots Coalition consisted of Ayres Vineyard, Belle Pente Vineyard & Winery, Beaux Frères Vineyards, Brick House Vineyards, Cameron Winery, Crowley Wines, Evesham Wood Winery, The Eyrie Vineyards, J. Christopher Wines, Patricia Green Cellars, and Westrey Wine Company.
**** If vine roots grow so deep, one might ask, what is the point of keeping the topsoil healthy? The answer is that the taproots grow deep in search

the plants to go dormant on the hottest days of the year; it doesn't push fruit to continue to produce sugars during heat spikes. It results in a naturally lower yield and smaller leaf canopy, which make for more concentrated fruit. In short, they posit, it creates vineyard conditions that are light on the water table and heavy on *terroir*.

Interestingly, "[n]either biodynamics nor organics address the deep-roots thing, not irrigating," notes Russ Raney, a Deep Roots Coalition founder and founding owner of Evesham Wood winery in the Willamette Valley.* "Our feeling is that if you are planting on a site that requires irrigation beyond what it takes to establish the vines, maybe you shouldn't have planted there." Indeed, Demeter USA's "Guidelines and Standards for the Farmer" apply to all types of farms and thus assume that some crops will require irrigation. A vineyard can be certified biodynamic while dousing its vines on a daily basis, provided that the irrigation water is "free of chemical contamination."

Another disconnect between Demeter certification standards and the general expectation for natural wines is that the first tier of certification—"made with biodynamic grapes"—allows for tinkering in the winery, such as acid adjustment and limited micro-oxygenation to encourage fermentation and smooth out tannins. But if the label states that what is inside the bottle is "biodynamic wine," then winemaking activities are limited to fining, filtering, some cold stabilization, barrel aging, and a quick shot of sulfur dioxide as a preservative. That is, no addition of any liquid ingredient that will later be detectable in the wine. In addition, plastic fermentation tanks are prohibited and corks must be recyclable.

Thus—the irrigation issue aside—Demeter "biodynamic wine" certification is just about as close as an American wine can get to proving itself to be *100 percent raisin.*

---

of water, while the remainder of the root mass remains near the surface, absorbing nutrients.

* Raney retired in 2010 after selling Evesham Wood to Erin and Jordan Nuccio, producers of Haden Fig. The Nuccios plan to continue to farm organically and refrain from irrigating Evesham Wood's vineyard, Le Puits Sec. Along with Brick House Vineyards, a certified-biodynamic DRC member, Evesham Wood is a member of Nicolas Joly's France-based biodynamic/natural wine trade group, *La Renaissance des Appellations* (referred to in English as "Return to Terroir").

But those who seek out biodynamic wines aren't pursuing merely simplicity or neo-naturalism. They want something more. They want wines with spirituality. Wines that aren't just *raisin*, but have, in fact, *raison d'être*. And this is where biodynamics deliver. Because, unlike any other farming system, BD has soul.

Doug Tunnell of Brick House Vineyards believes part of the recent spike of interest in biodynamic farming is due to the USDA's implementation of the National Organic Program in 2002. He describes a "disaffection" of organic growers with "the intervention of big business to try to get on the organic bandwagon." Says Tunnell, "There was a feeling that a high standard and a way of life for them was being co-opted by corporate interests" and biodynamic agriculture offered an alternative that was "more fundamental and more rigorous."

"On the part of consumers, there is a real surge in interest in something more real in life generally," notes Tunnell. "I see that in the people who come out here to taste. By the time they hit this barn or this part of the farm, you can just see the layers of shit peel away. They begin to just check out of whatever their world is, whatever they left behind when they started up that gravel road. When you start talking to people, and say, 'Here, get your hands dirty. Look at the worms,' it takes them back to their inner child and all those connections that, growing up in the twentieth and twenty-first century, we have just layered over."

Torn between technology and nostalgia, today's trendsetters take iPhone photos of their meals, then tweet about them, expressing their need for intimacy and sustenance using the most impersonal digital equipment. They are the post-post-modernist, affluenza-averse, introspective, old-school, alterno-hipsters. They consume the confessional blog, the mumblecore film, the freak-folk music captured on vinyl. They devour the sincere and self-deprecating, works of Dave Eggers and Spike Jonze. They sip single-origin coffees and send their children to Waldorf schools.* And they shop at the

---

* They might have flipped through the spring 2010 design edition of *T, The New York Times*' style magazine, and come across an article on a fascinating thinker and architect named Rudolf Steiner. "Today, design and architecture have become very focused on technology, removed from spiritual or social questions," a German museum curator laments in the

fiercely independent bottle shops and wine bars that have been cropping up lately in cities like New York and London and—it goes without saying—Portland, Oregon.

They might take a vacation in Burgundy, to soak in the thermal hot springs at Bourbon-Lancy, then sit in on a seminar at the Ecole du Vin et des Terroirs in Puligny-Montrachet. This institute, established by a group of biodynamic Burgundian luminaries such as Aubert de Villaine and Anne-Claude Leflaive, offers classes on Steiner-inspired disciplines such as "Listening to Wine," "Unlearning to Taste," and "Life of Soils and Wines of Terroir," as well as more abstract subjects like "The Language of Colors" and "Essential Oils."

It's all very deep.

One instructor at the Ecole focuses on the study of the "qualité d'un produit vivant," or energetic quality, of wine, using sensitive crystallization. This is a technique developed by the Steiner associate Ehrenfried Pfeiffer in which a substance in liquid form is combined with copper chloride and crystallized; when the liquid evaporates, the remaining crystalline structures are interpreted. When applying this method to human blood, Pfeiffer supposedly was able to detect the presence of cancer and other diseases. Today, it's fashionable to photograph sensitive crystallizations of wine made from conventionally farmed grapes and wine made from biodynamically farmed grapes. Comparing the two, the conventional images invariably look limpid and lifeless, like a smooth, sandy desert, while the biodynamic crystallizations tend to be dense and prickly, like a thicket of fine twigs. But while the images are compelling to look at, no one has yet explained why a very interesting pattern in a petri dish should be an indicator of plant or wine quality. "It is very variable," admits Randall Grahm, the charismatic leader of the California winemaking powerhouse Bonny Doon Vineyard, which has embraced biodynamic viticulture. "You need to look at a lot of images to come up with one that is consistent and actually reflective of the wine. It is a very imprecise science. It's an art, actually."

Marketing is an art, too, one that Bonny Doon has mastered: its labels have long been the holy grail of wine sales, with their spoofy titles, scribbled artwork, and irreverent humor copied (poorly) by also-rans the world over. Grahm left most of that behind him in 2006, when he sold off his hugely popular Big House and Cardinal Zin

article, sounding very much like a critic of contemporary agriculture.

brands, to focus on producing wines of quality rather than quantity. Since 2004, Grahm has been practicing biodynamic agriculture on his estate vineyards and contracting only with growers who are doing the same. He is hot in pursuit of *terroir*-driven—rather than dollar-driven—wines and redesigned his Ca' del Solo Estate labels to reflect this. Gone are the folksy, colorful woodblock-print labels. In are sensitive crystallizations against a sombre background. He's gone all spiritual on us.

Never mind that some of Grahm's wines are flavored with wood chips—an ingredient that fakes the effect of expensive French barrels—making him a not-exactly "natural" winemaker who happens to be using Demeter-certified grapes. (He has the guts, however, to list the chips on his back labels, where he discloses the presence of any additives used to make the wine.) For this denizen of Santa Cruz, California, sensitive crystallization is still important, as "just another lens through which you can understand your wine, like a mantra or a mudra."

One can't help but picture some 1970s New Ager, well versed in crystals (some say they were a source of power in the lost civilization of Atlantis), digging Grahm's words and his crystalline wine labels. It brings to mind the musings of the ponytailed consultant Alan York, who has called biodynamics "a methodology of arriving at a different state of consciousness that's not drug-induced." But the funny thing is, Bonny Doon's sensitive-crystallization labels are meant for mass consumption. Much like a psychedelic-looking fractal-image screensaver, the groovy crystalline images somehow fit within our contemporary aesthetic. Perhaps that's because the new millennium has ushered in a new New Age.

The first New Age movement percolated out of that esoteric circle that Steiner was so familiar with, most notably Madame Blavatsky's occultist theosophy and Jung's visions of an Age of Aquarius. Today, we associate the term with counterculture's New Age of Aquarius of the late 1960s and early 1970s, a cultural moment we tend to attribute to the ingestion of a few too many hallucinogens. But today's New Age is different. Because this time, it's egalitarian and it's ubiquitous. Flag-waving Republicans sport tattoos. Golfers wind down with hot-stone massages. Carnivores nosh on tofu bowls. Pot may be on the road to legalization. Christians hum Hindu chants at yoga studios. Surgeons dabble in the mind-body medicine of Deepak Chopra. Oprah watchers dig the spiritual guidance of Eckhart Tolle.

Hallucinogens have been replaced by wine; and today's stodgy wine collectors—who came of age in the '60s—think nothing of consulting astrological charts to determine the best day for popping the cork. When Randall Grahm tweets, "Mercury seemingly very retrograde," his nearly four hundred thousand wine-fan followers all nod knowingly. And when supermarkets and bottle shops throughout the United Kingdom plan wine tastings, they make sure to time them for astrologically favorable days.

In Burgundy and in Oregon, even those who don't claim to be practicing biodynamics consult the positions of the stars and planets in the heavens. Because pinot noir is the "heartbreak grape": even if the weather conditions at flowering, budbreak, and harvest favor this most delicate and sensitive variety, there is still the chance that the finest vintage from the finest producer can lack aroma and flavor on the day that you open a bottle. Like a lucky rabbit's foot, the timing of wine-related activities according to an astrological calendar might, the thinking goes, act as some sort of celestial insurance policy against "closed" wines. In the new New Age, consumers can dig this idea.

It's a bit wider of a leap to wrap one's mind around the idea of the "life force," or spiritual energy, that, according to Steiner and the New Agers, inhabits all living things. Biodynamic farmers sometimes speak of witnessing this etheric force when out in the field or the vineyard, when the plants seem to give off a colorful heavenly glow, or aura. Rudolf Steiner wrote and lectured with gravitas on the subject of gnomes.*

This is the point where biodynamic vintners' perceptions of themselves meet with the perceptions of the general public. People who have heard something about BD ask, "Isn't that like voodoo

---

* Deep Roots Coalition cofounder John Paul, the court jester of Oregon winemakers, once played a practical joke on fellow Coalition member and biodynamic practitioner Doug Tunnell: he snuck through Tunnell's Brick House Vineyards estate and placed garden gnomes in strategic places for Tunnell to find later. It goes without saying that while Paul farms and makes his Cameron wines naturally, he does not use biodynamic techniques.

farming or something?" Articles on the subject inevitably have "Voodoo" in the title. And just about every vintner I interviewed for this book jokingly described the practice as "voodoo." It's a word that just keeps popping up.

Those who don't understand biodynamics—and don't understand voodoo—use the term in reference to the preparations: the buried cow horns, the hanging stag's bladders, the cauldron of water, stirred methodically. They're thinking spells and incantations, black magic and dolls with pins in them. They're thinking Louisiana voodoo.

The biodynamic practitioners who describe their discipline as similar to voodoo appear to have a better sense of the religion as it is practiced traditionally in Haiti, where it is spelled "vodou": it's a nature-worship-based belief system. Its "fundamental principle is that everything—from humans to crocodiles to mango trees—has a spirit," as National Public Radio reporter John Burnett explained to curious Americans in the aftermath of the 2010 earthquake in Haiti. To the Haitians, this spiritual dimension is more important than the tangible dimension of everyday life as we experience it. Similarly, Steiner, following in the footsteps of Goethe and Plato, believed that everything on this earth has an ethereal archetype that is as important as the physical, spatial reality we see before us—if not more so. And then Steiner went so far as to apply these metaphysical ideas to tactile, tangible, science-based pursuits such as horticulture. And this is the fundamental difference between biodynamic farming and other forms of agriculture: it's spiritual.

It's a little-known fact that the biodynamic farming movement predated the organic. Sir Albert Howard's groundbreaking organic work, *An Agricultural Testament,* was first published in 1940, sixteen years after Steiner's lectures on biodynamic agriculture and twelve years after the introduction of Demeter certification.

During the Second World War, Ehrenfried Pfeiffer befriended J. I. Rodale, the American father of organic horticulture, and by all accounts, the two hit it off when it came to tangible concepts such as composting. Pfeiffer even developed a before-its-time municipal composting program in the 1950s in Oakland, California, that resembled the kitchen-waste disposal programs that some American cities are just beginning to implement today. But in the no-

nonsense 1950s, even as Steiner's lectures were being published and disseminated in the United States, it was Rodale's organic movement that took off. "They were both on the same page regarding the whole farming system. They were talking about composting and crop rotation and integrating animals. Pfeiffer had that more esoteric element and Rodale did not. That's where they split and organic became huge," says Demeter USA's executive director, Jim Fullmer. "It's really now just the past five years or so that biodynamic agriculture has been recognized."

Today, eccentrics with wild ideas aren't so much celebrated for being different as they are simply accepted into the mainstream. Consider *The Real Dirt on Farmer John*, the sleeper-hit documentary. Released in 2005, it chronicles the saga of an Illinois farmer/ raconteur who converts his property to organic and biodynamic agriculture as his neighbors are losing their acres to bankruptcy and corporate buyers. At first, John, a flamboyant ex-hippie with a taste for performance art and costumes, becomes the target of community hatred as the pastoral way of life is threatened by a big-ag economy. A scapegoat for the region's frustration, he is accused of hosting devil-worshipping orgies; a cabin on his farm burns to the ground under mysterious circumstances.

The film ends happily, but not because John has changed any of his way-out ways. He hasn't. Instead, his rural neighbors must acknowledge that his CSA concept was visionary. And his weirdness? Well, who isn't a little weird these days? Next to the latest Lady Gaga video, the image of Farmer John driving a tractor in a faux-leopard coat and pink boa looks positively tame.

In the wine world, the self-important Robert Parker, with his comb-over and air of gravitas, has come to represent fustiness and, worse, buffoonery. The new wine celebrities are young, idealistic, and borderline certifiable. People such as Gary Vaynerchuk, the multimedia multitasker who talks a mile a minute. And such as Monty Waldin, the young biodynamic proselytizer who moved from England to a vineyard in the south of France to make his own biodynamic wine and star in the six-part British television series "Château Monty" (2008).

Today's wine heroes don't wear ascots or smoke pipes. They're characters cut from the cloth of Randall Grahm and Farmer John.

They're poet-philosopher wine importers like Terry Theise and Kermit Lynch. They're young revolutionaries who like to think that by making or selling honest wine, they're sticking it to the Man. They're eccentrics like Alois Lageder, a fifth-generation winemaker in Italy's Alto Adige, whose winery is a museum for art installations and whose cellar is equipped to play slowed-down Bach and show slides to his wine as it ages in barrel. Lageder farms his grapes biodynamically, of course.

They're spiritual leaders-cum-vignerons like the celebrated Nicolas Joly, proprietor of Coulée de Serrant in the Loire Valley, author of numerous books on biodynamic wine, and, according to one American sommelier I spoke with, "bat-shit insane." "Sure, a lot of people think Nicolas is crazy," confirms the wine writer Alice Feiring. "The fact is, a lot of visionaries are crazy. That's just the way it is. The fact that he was able to create this worldwide passion is remarkable. Whether or not you think he is a brilliant winemaker is up to you."

They're the thinkers, the dreamers, the interpreters of stars and readers of crystals. They're farmers, neo-naturalists, locavores. They are anti-irrigators and *terroiristes*. They're nuts, and because they're nuts, their moment is now.

"For lunar, read loony. For certified, certifiable. That's no doubt how many sane people regard biodynamics," wrote British wine critic Anthony Rose recently in *The Independent Magazine*. The piece went on to praise biodynamic wines.

CHAPTER SIX

The Burgundians

His approach can be summed up in the classic saying of
the Burgundian vigneron: *Laissez le vin de se faire* (loosely
translated as "Let the wine make itself"), which sets down
the overarching rule that one must not disturb nature, that
man's role in the process is shepherd rather than master,
that if we work with and respect the vagaries of nature and
do our best to provide the proper conditions under which
the fruit grows and matures, the result will be splendid.
—Neal Rosenthal, *Reflections of a Wine Merchant*

One afternoon in late July 2001, hundreds of wine aficionados arrived
at Dillin Hall on the verdant, stately campus of Linfield College in
McMinnville, Oregon. The occasion was the annual International
Pinot Noir Celebration, a three-day wine-tasting extravaganza that
brings together collectors, critics, merchants, dealers, winery owners,
and vintners, all willing to pay close to one thousand dollars per
ticket.

For the seminar on this afternoon, tables placed end-to-end,
draped in white cloths, and set with tasting sheets, glasses, water
bottles, spit cups, and baskets of crackers filled the space with long
rows of white. The audience pushed in excitedly and competed for
the seats with the best views of the south side of the hall. On a small
stage here were two highly regarded American critics: the *Wine
Spectator* columnist and author Matt Kramer and Pierre-Antoine
Rovani, who was at that time Robert Parker's associate at *The
Wine Advocate*. In addition, Laurent Montalieu, the tall and affable
Bordeaux-born then-winemaker for WillaKenzie Estate in Yamhill,
stood by to act as translator.

Each of the gentlemen present was a lion in his own right, but
the crowds strained to see another, much smaller figure: a certain
Madame. She was none other than Lalou Bize-Leroy, the powerhouse
behind the wineries Domaine d'Auvenay and Domaine Leroy and
a shareholder in Domaine de la Romanée-Conti. A woman known
to wear a fur coat and heels in the cellar, Bize-Leroy has a birdlike
prettiness, her blonde hair often pulled back in a tight ponytail with

her bangs falling forward like a few loose feathers. A spritely sixty-nine at the time, she was a world-class mountain- and rock-climber, with a sinewy strength that was belied by her slight size.

Bize-Leroy has been a force to reckon with since the mid-1950s, when she took over the management of her father Henri's *négociant* house; she then made her name as a quality-driven winemaker during the 1980s, when she was codirector of Domaine de la Romanée-Conti (DRC), cementing its reputation for making the finest red wines in the world. By the late '80s, Leroy was acquiring her own vineyards with the help of investors; today, her Domaine Leroy in Vosne-Romanée and her Domaine D'Auvenay in St.-Romain produce astonishing wines that fetch prices of up to $2,500 a bottle. She stepped away from her role as *cogérante* at DRC in 1992 to focus on her own estates, but remains a shareholder.

Because of those high prices, and—it cannot be doubted—because of her gender, her diminutive size, her great strength, and her obsessive drive, Bize-Leroy had achieved a legendary wine-world status, exemplified by the nickname "La Tigresse." Thanks to her adherence since 1989 to the arcane practices of biodynamics, some have even seen her as a sorceress, an alchemist who can cast magic spells over hopeless vineyards and turn them into gold.

"I think Lalou's presence here was really like the oracle at Delphi coming down from the mountain," recalls Matt Kramer. "Lalou is more myth than woman. People were just fascinated to lay eyes on this near-mythological creature. It is hard to exaggerate how much Lalou is talked about and how little she is seen or heard."

Or, at their astronomical prices, how little her wines are tasted. But today, here were three 1999 Leroy grand crus (the Corton-Renardes, the Clos de la Roche, and the Latricières-Chambertin), waiting in elegant Riedel stemware for hundreds of lucky IPNC ticket-holders to try. "Her wines, of course, were breathtaking," Kramer continues. "It's one of those things where they are so goddamn breathtaking that there's nothing you can really identify in them. It's just not available to you. It's Burgundy at its pinnacle. People just tasted the wine and said, 'Oh, my god, this is really something incredible.'"

The audience sniffed and sipped and groaned, their eyes rolling back in their heads in ecstasy, as Bize-Leroy read her prepared remarks about the teachings of Rudolf Steiner and her application of biodynamics in her vineyards. Beginning in unintelligibly accented English, then switching to rapid French, she was completely unfathomable. It didn't matter. *Madame* was here—event-shy, she had in fact been ambushed into attending by her importer, Martine Saunier—and she was speaking, and the audience was tasting her wines and, well, just breathing in her presence.

"Biodynamics actually takes a back seat to the larger mystique of Lalou," says Kramer. "She could have said, 'I'm the daughter of Sun Ra,' and everyone would have believed her. She is capable of these things. She just occupies a realm unto herself."

Oregon wine is everything from the astounding Walla Walla syrahs of Christophe Baron's (biodynamic, in fact) Cayuse Vineyards to the fierce tempranillos of Earl and Hilda Jones, at their Abacela winery in Roseburg. It is cabernet sauvignon and zinfandel from the southern part of the state plus seas of plush, pleasing pinot gris from all over. But, above all else, in volume, in earnings, and in fame, Oregon wine is Willamette Valley pinot noir. And Willamette Valley pinot noir is a direct descendent of the pinot noir of Burgundy.*

Visit any Willamette Valley wine producer and you will find that there is an invisible thread that connects him or her to the motherland. At least a couple of local vignerons are Burgundian by birth. Others have studied in Beaune, married Burgundians, or worked in the great cellars of *Bourgogne*. Some simply open a different bottle of Burgundy every evening, wallow in its divine scent, and dream of creating something comparable on American soil.

And so, when Lalou Bize-Leroy came to the IPNC in 2001, those winemakers who were present may have claimed to have not heard a word she said. They may have protested that the way her wines tasted had no impact on their decisions to convert to biodynamics. And yet. Among those present at the IPNC that day in 2001 were representatives of four Willamette Valley wineries that would subsequently launch biodynamic programs, as well as one already biodynamic-certified. And the biodynamic study group that catalyzed biodynamic winegrowing in Oregon—see Chapter One—first began to meet in 2002.

---

* Willamette Valley vinetenders favor "Dijon clone" vines originally imported from Burgundy.

If you believe that Lalou Bize-Leroy is an enchantress, you might believe that she brought her special biodynamic mojo to Oregon that day in July and left the local wine community spellbound. Or, if you don't, simply consider these facts: For Willamette Valley winemakers, the pinnacles of achievement are the grand pinot noirs of Burgundy. And for Burgundians, the pinnacles of achievement are those grand cru vineyards that are farmed biodynamically.

Contrary to what American consumers might think, "There are thirty-eight hundred producers in Burgundy and only about three hundred of them are making really good wine; the other thirty-five hundred are sort of squeaking through on the reputation of their appellations, which is really unfortunate," confides the Oregon-based Burgundy importer Scott Wright, of Scott Paul Selections. "There is a sea of mediocre wine over there. It is just awful. You've got to kiss a lot of frogs to find the princes." Which explains why wine professionals are so acutely aware of the princes' pursuit and successful utilization of BD: "The percentage of people who do biodynamics in Burgundy is very small, but it's all the producers whom everybody knows: Romanée-Conti, Lafon, Leflaive, Lafarge. It's the crème de la crème," Wright explains.

Indeed, it's difficult to find a Burgundy grand cru these days that *isn't* incorporating biodynamics into its farming regime.* In addition to Domaines Leroy and d'Auvenay and those mentioned above, the list of top producers either dabbling in or embracing BD includes Domaine Dujac, Domaine Jean Grivot, Maison Louis Jadot, Maison Joseph Drouhin, Jacques-Frederick Mugnier, and Domaine Roulot.

To understand Bourgogne's recent interest in biodynamie, as the French refer to it, we must first travel back to the medieval era, when Benedictine and Cistercian monasteries began acquiring and farming the region's top vineyards. With their ample cellars and knowledgeable labor force, the monastic orders spent the better part of a millennium perfecting these sites. This era of devotional vinetending ended in 1789, when the revolutionary National Assembly abruptly confiscated and appropriated all church property.

---

* The Fork and Bottle Web site offers an unofficial list of estates embracing biodynamic farming practices: http://www.forkandbottle.com/wine/biodynamic_producers.htm

Now in private hands, the vineyards were soon entangled in Napoleonic legal changes, specifically the 1804 civil code that abolished primogeniture and mandated that property must be divided equally among all heirs. Its egalitarian spirit notwithstanding, this inheritance law ensured that Burgundy's historic large vineyards would be sliced into smaller and smaller parcels, sometimes to the point where, by the twentieth century, each shareholder owned just a single row of vines.

Then, in the 1970s and '80s, importers began arriving in France, eager to satisfy the demand of an increasingly sophisticated American market for gourmet cuisine and the appropriate wine to accompany it. Seeing an economic opportunity, absentee *domaine* owners returned to their estates and began to buy back their vines from the complex networks of neighbors and distant relations who controlled them. In many cases, they found parcels of land that had been neglected or over-treated with chemicals by sharecroppers, who had been more focused on harvesting fruit of quantity than quality. As *domaine* owners struggled to improve the health of their bedraggled vines, they began to look critically at the farming methods that had left their land in such a state.

According to biodynamic wine author and expert Monty Waldin, BD slowly began to gain traction at this time, with five producers committing themselves to *biodynamie* by 1986. But the watershed event may have been a November 1989 viticulture conference in Chalon-sur-Saône at which the French National Institute for Agricultural Research agronomist and soil microbiologist Claude Bourguignon presented the results of two hundred measurements of biological soil activity in vineyards of the Côte de Nuits and Côte de Beaune. Bourguignon's declaration that the soils of Burgundy harbored less microbial life than the sands of the Sahara Desert made local newspapers and sent the vigneron community reeling.

In addition to identifying the problem, Bourguignon simultaneously had been researching a solution: since 1979, he and his biochemist/oenologist wife Lydia, along with professor Max Léglise, at that time the director of the oenological research center at Beaune, had been spending their free time testing biodynamic viticultural practices on a vineyard in Beaujolais.

Encouraged by the increased microbial activity they were finding in their experimental plots—and their work starting in 1984 for a certain Loire Valley client named Nicolas Joly*—Claude and Lydia started up their own independent laboratory and consulting service in 1989. That same year, they conducted soil analyses of Leroy's Romanée-Saint-Vivant and Clos de la Roche vineyards. Alarmed by what they found, they encouraged Lalou Bize-Leroy to convert to biodynamic agriculture. Leroy did so immediately, and the rest, as they say, is history. Since then, the Bourguignons' LAMS (an acronym for, in English, Laboratory for Microbial Soil Analysis) has conducted more than six thousand complete soil analyses all over the globe and the couple has consulted for Burgundy's top estates.

The most notable of these may be Domaine Leflaive. Starting in 1990, Anne-Claude Leflaive—eager to try *biodynamie* but needing to convince the rest of her family of the value of this unorthodox style of viticulture—contracted the Bourguignons to launch a thorough comparison study of the effects of organic and biodynamic agriculture on her Le Clavoillon and Bâtard-Montrachet vineyards. Word got around, and soon neighboring vignerons and international critics were showing up to observe Leflaive's experimental plots and blind-taste her organic and biodynamic wines side by side. By 1998, Leflaive had converted her family's entire estate to biodynamic, and her straightforward approach had convinced more than a few skeptics to take BD seriously.

The same year, the Bourguignons began biodynamic trials for the hallowed king of all Burgundy. Domaine de la Romanée-Conti had been farmed organically since 1986, but *cogérante* Aubert de Villaine was looking for additional ways to fight problems such as mildew in the vineyards, and felt that biodynamics might offer alternatives. With the Bourguignons' assistance, he converted sections of the Romanée-Conti, La Tâche, and Richebourg vineyards to BD. Apart from a devastating attack of *court-noué* (fanleaf virus) in 2000 that required chemical intervention, the experiments went well enough that de Villaine felt confident about converting the entire estate in 2008. Although he has been hesitant to discuss this change in

---

* As mentioned in previous chapters, Nicolas Joly is proprietor of Coulée de Serrant in the Loire Valley, author of numerous books on biodynamic wine and founder of the natural wine group Return to Terroir.

viticulture techniques—writing in a recent e-mail, "*Que cela fait partie de notre philosophie mais n'en est pas du tout le point le plus important*"*— he now describes the estate as "entirely biodynamic."

Since 2000, de Villaine has been part of a select group of Burgundian producers who meet regularly with the consultant Pierre Masson to discuss techniques, compare trials, and observe results in the vineyard. They include Emmanuel Giboulot, Thibault Huber, Frédéric Lafarge, Dominique Lafon, Anne-Claude Leflaive, Didier Montchovet, Pierre Morey, Eric de Suremain, Jean-Louis Trapet, and Philippe Drouhin. Estates manager of the famed Burgundian *négociant* empire, Maison Joseph Drouhin, Philippe Drouhin is another strand in the web that connects Burgundy to Oregon. In 1987, his family founded Domaine Drouhin Oregon (or DDO, as Oregon wine-industry insiders call it) in Dayton, where he oversees viticulture and his sister Véronique Drouhin-Boss manages oenology; Philippe considers the Willamette Valley to be his second home.

"Philippe is a very un-touchy-feely kind of guy," confides the importer Scott Wright, who was previously the managing director of Domaine Drouhin Oregon. "He is the kind of guy who says, 'I don't believe in all of this romantic bullshit. I want to see cold, hard facts.'" So when Wright visited the company's Burgundy properties in 2001, he was astonished to find Drouhin practicing biodynamic viticulture. But Drouhin showed Wright vineyard parcels, previously diseased and struggling, that were now thriving and producing balanced fruit. "To see someone like that who is a hard-core black-and-white science guy embrace it is what really turned me onto it," recalls Wright, who admits to being something of a romantic himself. And so, when he left Domaine Drouhin Oregon to start his own label, Scott Paul Wines in Carlton, Wright set about farming his own Willamette Valley vineyard blocks according to biodynamic principles.

Here was a direct example of what had been going on for the past decade indirectly: Pinot noir producers in the Willamette Valley— and, indeed, all over the world—had taken note of the change underway in Burgundy. And they had begun to follow suit.

But as enthused as he is about it, Wright—who once ruined a pair of new shoes tromping through the Clos des Chênes vineyard

---

* Roughly, "That's part of our philosophy, but not the most important part."

in Volnay with the Masson study group as its members inspected Frédéric Lafarge's spray experiments—admits that biodynamic agriculture fits with the French psyche better than it does with the American. "Americans tend to plant a block of 777, a block of 115, a block of Pommard.* Everything is very regimented. In Burgundy, they plant as many clones as they possibly can and mix them up to get the full expression of the site," Wright reflects. "It is very much a philosophy and a work in progress for them. They are trying, experimenting, constantly playing around to see what different results they get. Whereas Americans are more like, 'OK, what's the prescription?'"

(Perhaps this disconnect explains why, so many years after adopting *biodynamie* in Burgundy, Drouhin still has not done so in Oregon. After two decades of contracting out his viticultural work here, Drouhin hired an in-house viticultural team in late 2007. According to Philippe Drouhin, DDO's vineyards have since achieved LIVE status and are working toward organic certification, with the long-term goal of converting to biodynamic.)

No one doubts that there are profound cultural differences between France and the U.S., but we seldom think about how these differences express themselves in the allegedly neutral realm of science. Take medicine, for example. As *The New Yorker* staff writer Adam Gopnik has pointed out, French obstetricians don chic black ensembles instead of white lab coats and exhort their patients to drink red wine: "In New York pregnancy is a ward in the house of medicine; in Paris it is a chapter in a sentimental education, a strange consequence of the pleasures of the body." And so it is with horticulture. Burgundian vignerons tend to rely on their instincts rather than spectrometer readings; they decide when to harvest by tasting fruit, sniffing the breeze, and considering the appearance of their vines rather than by measuring Brix levels. As Wright points out, they are carrying on holistic traditions born more than a thousand years ago, when Burgundian monks spent countless hours in their vineyards, experiencing spiritual epiphanies among the vines.

---

* 777, 115, and Pommard are grapevine clones. Because each clone reacts differently to climatic conditions and produces fruit with different characteristics, most vinetenders believe that clonal diversity is key to ensuring a successful pinot noir harvest.

Trained in Burgundy but born in Canada, Isabelle Meunier has made wine on both old- and new-world estates, and thus has a nuanced perspective on this cultural divide. Today she is winemaker for the Oregon arm of Evening Land Vineyards, an American venture under the consulting eye of Dominique Lafon, of the acclaimed Meursault estate Domaine des Comtes Lafon. "As Dominique says, the gamay noir gets harvested when you shake the vine and the clusters fall off," Meunier says. "Those are old traditions that have been happening for hundreds of years. You can check with your instruments to confirm your methods, but these methods are proven and they are true. Over here, we rely a lot on our machines and don't use our instincts as much, partly because we don't have as much history."

Evening Land Vineyards is yet another thread pulling the Willamette Valley and Burgundy closer together. The New York-financed venture includes two top-rated vineyards in California, Occidental and Odyssey, as well as Seven Springs Vineyard in the Willamette Valley. But Dominique Lafon prefers not to work with what he considers to be overripe California fruit. "To me, Oregon is closer to what we are trying to achieve in Burgundy, in terms of style of wine," he once confided.

But Lafon *is*, however, willing to travel to the Willamette Valley's Eola Hills a few times each year to oversee the harvest and winemaking of the Seven Springs Vineyard fruit. After signing on to the project in 2007, Lafon hired as his on-site oenologist Burgundy-trained Meunier, who had experience making biodynamic wines both in France and in Canada, and the two have convinced vineyard manager Stirling Fox to convert Seven Springs Vineyard to biodynamic.

Lafon became interested in BD as early as 1991 or 1992, on a visit to the biodynamic Domaine Huët in the Loire Valley. Witnessing the success of Leflaive's side-by-side studies was enough to inspire him to commence trials of his own, beginning in 1995. "My guys in the vineyard liked it better," Lafon recalls about the initial biodynamic experiments. "They said the vines were easier to work with, in terms of growth. They also felt better in terms of health." (Ironically, Lafon is one of those Frenchmen who reeks of cigarette smoke; one hopes that the health of his vines will rub off on their proprietor.) As more and more visitors, consultants, friends, and neighbors began to compliment him on how vibrant his vineyards looked, Lafon decided to convert all of his holdings in 1998.

Fast forward to 2008. Just as Evening Land's spare winery in Salem's industrial district was nearing completion in time for harvest, Lafon traveled to Oregon to present a Domaine des Comtes Lafon Meursault and four vintages of Volnay at the IPNC, in another show-stopping Dillin Hall seminar for a phalanx of starry-eyed attendees.

The perpetually perky Master Sommelier and popular wine educator Andrea Robinson introduced Lafon. Waggish Master of Wine and Burgundy expert Jasper Morris moderated. Everyone—excepting, perhaps, Robinson, who sounded chipper as ever—was hung over. "We fed him some port last night to make sure he would be in good form today," chortled Morris in his booming baritone, indicating a slouching Lafon.

"I'm not going to fall apart," retorted Lafon. Still boyish at forty-nine, with floppy sandy-brown hair, hooded eyes, and a voice in the tenor register of a well-tuned violin, the strapping aristocrat turned to the audience. "The first thing to know about sustainability is that wines produced this way don't hurt you, even if you drink a little bit too much," he said with a sheepish smile. Lafon's Oregonian friends—among them vintners from Adelsheim Vineyard, Ponzi Vineyards, and Domaine Drouhin Oregon—chuckled. Then the crowd listened attentively as Lafon recalled taking the reins at his family business in the mid-1980s: assuming former sharecroppers' contracts, he had found the vine health to be deteriorating. This had led to his discovery and integration of biodynamic viticulture.

He was stopped short by an audience member, who raised his hand to ask what was this "Vitamin E" product that everyone was talking about?

"No, BEE-OH-DEE-NAH-MEE. It's a system. I don't use products!" Lafon retorted. "If I get into that and give you a lecture on *biodynamie*, it's going to take a long, long time. It's as simple as it is complicated," he went on with a shrug.

So much for progress. Seven years after the seminar at which Lalou Bize-Leroy had introduced her rare biodynamic wines to the international audience of connoisseurs at the IPNC, some of the most discerning wine lovers in the world still hadn't clued into this style of viticulture. But, unlike in 2001, every *winegrower* in the room knew exactly what the great French vigneron was talking about. "I was probably one of the converted," admits the Willamette Valley winemaker Laurent Montalieu, who stood onstage with the panelists both years and now practices biodynamic viticulture at his own estate. "I think what has cemented the relationship between Burgundy and Oregon is the IPNC. That is a very special connection," reflects Montalieu. "The Burgundians have been very, very generous in sharing everything they have discovered."

CHAPTER SEVEN

The Oregonians

Oregon seemed inhabited, in my limited view, by folks
who often were of a stubbornly independent and even
renegade character, never quite convinced of the perceived
wisdoms and blessings of the wider world. You had every
variety of the dreamy and discontented ... all remaining
in or coming to Oregon to seek or pursue some insistent,
uncontrollable and potentially soul-wrecking passion.
If writing is one such passion, surely winemaking must
be another, especially for anyone trying to turn a fickle,
delicate, sometimes inscrutable grape like Pinot Noir into
the wine that Oregon is famous for.

—Chang-rae Lee, *Food & Wine*

In the city that harbors its fair share of shamans, aromatherapists,
masseuses, ayurvedic healers, and herbalists, in the city that is
home to the National College of Natural Medicine, the Institute
for Traditional Medicine, and Artemisia, the Association for
Anthroposophic Health Professionals, it was not unusual in 1990—
before much of the rest of the nation was hip to such things—that a
psychiatrist and MD named Robert Gross should become interested
in alternative healing.

After years of studying and working in the area of chronic pain,
Dr. Gross had discovered acupuncture to be an effective treatment
for his patients at Oregon Health & Science University hospital in the
early 1990s. Upon completing a medical acupuncture certification,
he became proficient in the practice and began studying other
alternative therapies: homeopathic, and then anthroposophic,
medicine. Researching the history of these traditional remedies,
Gross became intrigued by the figure of the sixteenth-century Swiss
physician and pioneering botanist Phillippus Aureolus Theophrastus
Bombastus von Hohenheim, also known—thankfully—by the nickname Paracelsus. Widely regarded as the inventor of toxicology,
Paracelsus has also been called an alchemist, astrologer, and
occultist, and the forefather of homeopathic medicine. "He came up
with the idea of 'like cures like': that is, if you find a substance that

in large amounts causes symptoms that look like the disease, then infinitesimal amounts of that substance will cure it. That created the backbone of homeopathy," explains Gross. It also contributed to the theoretical framework for one of Western medicine's greatest breakthroughs: the vaccine.

The more Gross read, the more he became intrigued by the medical scholars of centuries past. He followed his study of Paracelsus with one of Samuel Hahnemann, the nineteenth-century German who invented homeopathy. From there, the next obvious figure to investigate was an Austrian named Rudolf Steiner. Gross does not claim to be an expert in anthroposophic medicine—his practice doesn't go beyond acupuncture and homeopathy—but his personal reading of Steiner took him in a different direction. As he studied Steiner's works, he began to think not so much of his patients but of his vineyard properties, the first of which he had planted in 1978.

"Steiner grabbed onto the same idea Paracelsus and Hahnemann had with natural substances. The substances that he used—like dandelion, oak, quartz, or silica—are all homeopathic substances. His dilution process in making the preps is what you'd call 'succussion' in homeopathy," reflects Gross. "I was in two places at the same time, with both coming together and converging in my head. I'm not sure that there are any other doctors who have crossed the agricultural and medical lines in this way."

And so, in 1999, after four years of biodynamic cultivation under the tutelage of the consultant Alan York, Cooper Mountain Vineyards became the first Demeter-certified vineyard property in Oregon and the fourth in the United States. The news didn't shake the Oregon winegrowing culture at the time. In fact, it barely made a ripple. "I don't think there was any reaction at all," recalls the prominent Portland-based wine critic Matt Kramer. "It would have been different, for example, if Eyrie* had become biodynamic. But Cooper Mountain plays no real role in the larger Oregon wine ethos."

Gross was a physician who happened to own a winery, not a full-time winegrower. Although he has been instrumental in bringing influential speakers like the power consultant Alan York and the

---

* The Eyrie Vineyards, established in 1966, is credited with establishing the Willamette Valley pinot noir industry and remains one of the most respected wineries in the region today.

French biodynamic evangelist Nicolas Joly to Oregon wine country, he admits to being shy about reaching out to his biodynamic cohorts in the industry. Plus, he has a reputation in the industry for gruff intractability. "Bob kind of sits up on his hill and does whatever he feels like doing," observes one Oregon winemaker.

A past chair of the American Academy of Medical Acupuncture, Gross today treats his patients with a combination of conventional and alternative therapies, more on the model of a European doctor than an American physician. And he has radically altered his farming practices as a result of his quest to ease human suffering.

Dr. Robert Gross may not be the most notable of the Oregon biodynamic vintners, but he was the first. In a time when "biodynamic" was an unfamiliar term for American wine consumers, Gross made a move that showed a reckless disregard for the bottom line. In his determination to stubbornly slog down an unbeaten path, he was acting the role of the quintessential Oregonian.

Like people who go into politics, Oregonians tend to be tough, hard-headed, and slightly unhinged. We are quick to claim the tenacious and foolhardy Meriwether Lewis and William Clark as two of our own. Anyone who is crazy enough to trudge westward for eighteen months only to spend a long, miserable winter in the rain at Fort Clatsop, then turn right around and trek back, gets an honorary citizenship. Oregonians are people like James Beard, Matt Groening, Ken Kesey, and Gus Van Sant. They are leaders in their respective fields, certainly, but they travel their own offbeat paths to success. They have little interest in what the world at large thinks about what they are doing. "I think that they believe in all the things everyone else believes in; they just take a different approach to getting there," reflects Charles Heying, a professor of Urban Studies and Planning at Portland State University and the author of *Brew to Bikes: Portland's Artisan Economy*. "They do it the zen way: don't aim at the target and the target will come to you." As the Portland ad firm so famously put it for the Beaverton shoe company, they Just Do It.

From the late 1960s through the mid 1970s, Oregon had an intractable governor named Tom McCall who, having grown up on a ranch in the central part of the state, had a near-religious regard for the sanctity of open spaces. He didn't care that Republicans aren't supposed to be environmentalists. In a few years' time, McCall kick-started container recycling in the United States by enacting the nation's first comprehensive bottle bill and helped to pass pioneering land-use laws that are still the most progressive in the nation: the state's entire coastline was preserved for public access, and urban-growth boundaries limit land use outside of densely populated areas.

Nearly three decades after McCall's death, Oregon remains a place where many people are such fierce, rabid environmentalists that they often show a reckless disregard for the bottom line. We treasure our bicycle lanes here in the western part of the state because we like to get places the slow, hard way, even when it rains every day for eight months. And if Oregon is green, the Oregon wine industry is the glowing chartreuse of a vineyard in springtime. It's—as Robert Gross would put it—"Ore-ganic."

In a state known for its sustainable building designs, some of the most notable are wineries, where solar panels sprout up as quickly as grape shoots. In the Willamette Valley, Stoller and Sokol-Blosser were the first LEED gold- and silver-certified wineries, respectively, in the nation.

In 2009, the Oregon Wine Board launched the Oregon Certified Sustainable brand, an umbrella program that brings all eco-labels under one easy-to-recognize logo. It was a necessity in a state where the eagerness to get certified is only matched by the number of possible independent third-party certifications, including LIVE, Salmon-Safe, Vinea, Food Alliance, organic, and biodynamic. That same year, the Board partnered with the Oregon Environmental Council to mount the Carbon Neutral Challenge. By 2010, the number of carbon-neutral wineries numbered fourteen in Oregon, compared with only two others in the rest of the nation.

These initiatives have been groundbreaking, certainly. But in Oregon, they have been par for the course. This is the state that is home to the oldest BD club in the country: the Oregon Biodynamic Group, an informal organization of gardeners and farmers, has been meeting since 1975 to collectively make preparations and discuss Rudolf Steiner's teachings. This is the state that is home to Oregon Tilth, which drafted the nation's first standards and procedures for organic production in 1982 and today remains the second-most prominent certifier nationwide for the National Organic Program, as well as operating in six foreign countries. Six miles west of Tilth's home base of Corvallis are the national headquarters of Demeter USA, the biodynamic-certification organization, in Philomath. The Biodynamic Farming and Gardening Association, which publishes the quarterly journal *Biodynamics*, is headquartered another thirty-two miles south of Demeter, in Junction City.

Demeter USA executive director Jim Fullmer started out as an Oregon Tilth certifier, traveling all over the globe to visit and accredit organic farms. "Oregon is a happening place," says Fullmer when I ask him why so many eco-agricultural institutions are based here. "It always has been when it comes to the ecological movement. Look at all the organizations based in Portland alone. There's a lot of sympathy for it in this gorgeous, once-upon-a-time rainforest."

Indeed, out-of-state visitors are always sniffing our clean, crisp air. And then, inevitably, remarking about how green it is here. Not just figuratively, but literally, green. In the Willamette Valley, as I've stated previously, it pretty much rains eight months out of each year. Then comes the summer heat that can occasionally surpass 100 degrees Fahrenheit. The result of such a potent combination of water and light is a place where gardens spill out onto sidewalks, where trees appear to spring from rocks,* and where roses grow as large as dinner plates. It is shockingly easy to cultivate things here, which is why it is so easy to be an organic gardener here. And it's why biodynamic agriculture, irrational as it may appear, is not all that nonsensical of a leap for an Oregon vinetender to make.

One summer at the International Pinot Noir Celebration, during a visit to WillaKenzie Estate in Yamhill, I got into a conversation about BD with Maxime Rion, a Burgundian vigneron from Nuits-Saint-Georges. He told me he and his family had been conducting some biodynamic tests in their vineyards, but then he looked around him. It was one of those summer days in the Willamette Valley when a vineyard can feel like a sauna; heat radiated from the gravel road and the bright-green vines seemed to glisten and steam. "Here everybody should do it—it's so easy!" he exclaimed. "In Burgundy it is so rainy, even in the summer. There is too much pressure from rot and disease."

---

* The Columbia River and Oregon coast are both famous for their massive tree-covered rock formations.

Sheila Nicholas was born in Iran, the child of a globe-trotting British diplomat. The owner, with her husband Nick, of Anam Cara Cellars, northeast of Newberg, has a similar outsider's perspective. When she remarks, "The air is so pure here!" she speaks from the perspective of someone who has breathed the air in a wide variety of places. Before moving to the Willamette Valley, Nicholas lived in the Napa Valley, working in public relations for the Terlato Wine Group, which exposed her to influential biodynamic vintners such as Michel Chapoutier and Nicolas Joly. In 2001, shortly after arriving in Oregon, she joined the biodynamic study group that consisted of Moe Momtazi and Jimi Brooks, Doug Tunnell and others. Today, her vineyard is certified LIVE, but Nicholas also incorporates some biodynamic practices into her viticulture.

When Nicholas looks around her vineyard, surrounded as it is by grasses, wildflowers, and tree-covered hillsides that are hemmed in by fog on many mornings, she is reminded of her childhood summers spent in the Scottish countryside. The biodynamicist's toolkit of local flora—nettles, dandelions, horsetail, and oak in the wild; chamomile, yarrow, and valerian in the garden—ubiquitous in the Pacific Northwest of the United States, also play prominent parts in the Scottish landscape. "I would climb those hills and feel this amazing sense of spirituality," Nicholas recalls. "I feel a lot of that here, as well. I think there is something about vines, the depth of their roots and how they translate the soil. Soon after we prune, the vines are still bringing up water. If you ever taste the water on those cuts, it's the most clear, translucent ethereal liquid."

Producing two thousand cases of wine each year, Anam Cara Cellars is your typical Oregon outfit: boutique and high-end, with a range that includes a $22 riesling and a couple of $65 pinot noirs. "I don't know of anybody who is farming their vineyard conventionally anywhere in Oregon," says Nicholas when I ask her why biodynamic viticulture is so prevalent here. "And the smaller the vineyard, the more likely they are to go BD because one person can control it."

The wine writer and critic Matt Kramer agrees: "It's here in the culture; it's here in the air. The very DNA of Oregon winegrowing is sympathetic to this non-interventionist, naturalist, small-scale form of farming and winemaking, whether biodynamic or any other form," he says. "Biodynamics, because of its French origins, happened to have attracted more adherents."

The typical Oregon winemaking operation produces five thousand cases of wine or fewer annually. Oregon's single largest producer, King Estate in Eugene, makes two hundred thousand cases—small potatoes in the grand scheme of the global wine market. (By contrast, E&J Gallo, California's biggest player, churns out seventy-five million cases of wine annually.) And, even though it's not technically biodynamic, King Estate comes awfully close. The property encompasses 1,033 certified organic acres of grasslands, forest, fruit, and vegetable gardens, and the world's largest contiguous organic vineyard (470 acres).* Fertilization at King Estate comes from a thousand tons of compost a year. Wild turkeys roam the property; rehabilitated birds of prey from the Cascade Raptor Center keep the vineyard gopher-free; and a flock of between three hundred fifty and five hundred fifty sheep mows the weeds and grass that grow between the vine rows. If this is what large-scale corporate winegrowing looks like in Oregon, it should come as no surprise that the boutique producers are increasingly busy burying cow horns. In this state, the bar for "artisanal" is high.

If they are anything, Oregon vintners are collaborative. They have the highly unusual habit of meeting annually at the Steamboat Inn** to analyze and criticize each other's very worst wines, with the goal of collectively determining what went wrong and how everyone could avoid such a problem in the future.

"We're more willing to share our information and our equipment and help each other," affirms Bill Hanson, winegrower for Libra Wine in the Willamette Valley's Yamhill-Carlton District. "If anyone makes a bad wine, it affects all of our reputations. We are not trying to compete against each other as much as trying to build Brand Oregon.

---

* "From 10,000 B.C. to 1945 A.D. All Agriculture Was Organic," reads King Estate's latest ad campaign. "We owe the future."

** The Steamboat Inn is situated on a picturesque stretch of the Umpqua River, a world-class destination for those in pursuit of Coho, steelhead, Chinook, and rainbow trout. Oregon winemakers do not under any circumstances divulge their best casting spots. Trade secrets, however, are fair game. Alas, journalists are not invited to the Steamboat Pinot Noir Conference: www.steamboatpinot.com

There is camaraderie here." That camaraderie is evident in Hanson's dual identity: On weekdays, he's the assistant winemaker and sales and marketing manager for Panther Creek Cellars in McMinnville. Nights and weekends, he runs Libra, using Panther Creek's facility to make his wine. You could say he's poaching from the competition, but in Oregon, this is standard practice. Large wineries are breeding grounds for new talent; large winery owners are the de facto patrons for up-and-coming winemakers. In fact, Libra isn't the only offshoot of Panther Creek Cellars; Panther Creek's chief winemaker, Michael Stevenson, also has his own side label, Stevenson Barrie. In effect, this means that there are three wineries under one roof, all making high-end pinot noir, choosing to help one another rather than to compete.

In the typically familial manner of the Oregon wine scene, Libra's Bill Hanson fell in love with the biodynamic fruit from Momtazi Vineyard and introduced it to the Panther Creek program at the same time that he began contracting it for his own Libra label. Now, Michael Stevenson uses Momtazi fruit for his Stevenson-Barrie wines, as well. All is shared within the family. "They delivered beautiful fruit in 2007, in about as difficult as a season can be.* There is a brightness to it; it's so alive. This wine has so much nerve," Hanson says admiringly as he swirls a glass of his Libra Momtazi Vineyard Pinot Noir.

Despite the fact that he's a mere "weekend warrior" at his own Libra vineyard, Hanson composts religiously and applies biodynamic preps when he can. Being a neon-green Oregonian, he irrigates his newly established vineyard with rainwater, collected in twenty-two massive thirty-five-hundred-gallon tanks parked beside his house. Still, "I'm really just dabbling in biodynamics," he says. "You really have to know what you're doing. It's like bringing a knife to a gunfight." Moe Momtazi, he adds, is the true master: "He is like a samurai warrior who doesn't need a gun."

Hanson's Libra label depicts the Greek goddesses Persephone and Demeter, frozen in a moment of an intricate dance. Holding hands, they both lean backwards, one leg raised, counterbalancing one another, active yet static. Their pose is one of momentary insanity: If one loosens her grip, both deities will fall. But they don't let go. They

---

* 2007 was a notoriously cold and rainy vintage in the Willamette Valley.

symbolize a perfectly balanced wine, but also a strong and graceful collaboration.

Whatever you thought of his wines, it was hard not to notice Jimi Brooks, a restless world traveler and a charming rogue with an endless curiosity and predilections for bourbon, Moto Guzzi motorcycles, Unimog lorries, and vintage Land Rovers. When Corby Kummer, a senior editor at *The Atlantic*, traveled to Oregon in 2003 to report on the wine scene here, he ended up focusing his article on this compelling young vintner's struggle to build his brand.

Lacking a formal pedagogy, biodynamic agriculture is a practice that spreads largely by word of mouth. This is where the effect of influencers like Jimi Brooks cannot be underestimated. Biodynamic agriculture would never have been so widely adopted without the charisma of Rudolf Steiner; and it could be argued that biodynamic viticulture would not be so widely accepted in Oregon as it is today without the early enthusiasm of Brooks, the biodynamic winemaker and vineyard manager for Maysara and its Momtazi Vineyard as well as for his own Brooks label.

In 2004, when Brooks died suddenly of an aortic aneurism at the age of thirty-eight, it was harvest time. A dozen local vintners—his wine-community family—immediately leapt to the aid of the Brooks label, bringing the fruit in and vinifying it on Brooks's behalf that first year while they mourned; the group continues to offer advice and support to Jimi's sister, Janie Brooks Heuck, and friend and now-winemaker Chris Williams, as they continue to run the label on behalf of Jimi's teenage son, Pascal.

Had he lived longer, one could imagine Brooks becoming the Oregon equivalent of California's Mike Benziger, a tireless sixty-ish promoter of BD with a puckishly youthful appearance and energy, as though all that time spent among homeopathic cures for grapevines were some sort of fountain of youth.

"Jimi Brooks was a catalyst for this. He was instrumental in getting this thing going," says the winemaker Jay Somers when I ask him why he decided to practice biodynamics in his vineyards and winery. "I became familiar with what it was all about through him. He was a very charismatic guy, very passionate." Like Brooks—who grew up in Portland and attended Linfield College in McMinnville—

Somers is a lifelong local; the two became friends in 1989, when, fresh out of college, they began working in first the brewing, then the winemaking, industries together. Somers's Swiss ancestors arrived here after a hardscrabble covered-wagon journey along the Oregon Trail back in the early 1800s; Somers grew up in Medford and Portland and attended the University of Oregon. He calls himself "A diehard pinko environmentalist and a fiscal conservative." In short, he is your typical Oregonian.

Back in February of 2003, I visited Somers to talk about riesling. Scheduling an appointment with this guy had been a matter of tracking him down on his cell phone and making out what he was yelling about over the sound of a tractor or forklift or van or, incongruously, a drum set. He worked in a drippy, unheated, open-ended barn. As is the custom in Oregon, his employer allowed Somers to use the space to produce his own label, J. Christopher.

Somers's hands were raw and rough from exposure. I could see his breath when he exhaled. His Carhartts were dirty. His nose was red. So many fermentation tanks and barrels were crammed into the barn that there was barely enough space to walk between them; tarps kept the rain from beating down on the crushpad. One had a sense that the whole operation could come toppling down at any moment and flatten the fellow.

Like so many Willamette Valley vintners, when he wasn't out in the vineyard pulling leaves or in his precariously patched-together winery rolling barrels around, Somers was behind the wheel of his van, driving to Portland wine shops to deliver his fermented juice himself. He wore his hair long and curly, and claimed that winemaking was just a day job; his real love was playing guitar for the jazz-funk band JJ and the Rhythm Dogs (he's since moved on to a band called Poncho Luxurio).

"The part of this I like is making the wine," Somers told me in 2003. "The part I don't like is selling it. I'm not willing to go out there and try to convince people that my riesling should sell for $30 a bottle so I can make a good margin on it. It's more about surviving than making money."

I thought, Oh, dear.

At that time, Somers's wines, under the Holloran and J. Christopher labels, were available in five states in addition to Oregon. I asked if he had ever considered submitting some bottles to any national critics. "I'm not sending my wines to anyone," he retorted. "If Parker wants to taste my wines, he can just come here and visit me and see what we're doing." The tarps dripped. He blew his nose.

I thought, Oh, dear.

Somers's wines were breathtakingly beautiful in 2003. They still are. Over the years, I have put them into paper-bagged blind lineups again and again, and whether I'm alone or with a group of colleagues, we always select his as our favorites. They are clean. But more than that, they are elegant. They are emotionally stirring.

So, here's the funny thing: in 2002, Ernst Loosen, the most successful winegrower in all of Germany—he runs his own family's two-century-old estate on the Mosel River in addition to another winery in the Pfalz, and also created Chateau Ste. Michelle's marquis riesling, Eroica—happened to be visiting McMinnville for the International Pinot Noir Celebration, where he happened to taste this irreverent Oregon winemaker's wines. The famous German vintner (who also, incidentally, has an impressive head of curls) was smitten by Somers's gamine yet powerful pinots. He invited the wild-haired guitar player to work alongside him in Bernkastel in November of 2004; while drinking a great deal of Burgundy from the German's cellar, the two hatched a plan to make pinot noir together. In 2006, Harvey Steiman of the *Wine Spectator* wrote about the collaboration on his blog. The unknown winemaker from Oregon was, suddenly, known.

Meanwhile, in 2005, Somers had begun farming biodynamically, inspired by his friend Jimi Brooks. He had seen how much healthier the five BD Brooks acres of the Eola Hills Vineyard had been, incongruously producing more fruit and higher sugars, as well as less botrytis, than his conventionally farmed acre of the same property. He remembered the passionate exhortations of his recently deceased friend. And he wanted to save the 1972 riesling vines at Holloran's Le Pavillon vineyard, which were losing vigor and looking sickly. "To be frank, I can't really say the wines are way better, but we have a vineyard that is not slowly dying; it is revitalized and still making great wine," says Somers today. "My next decision as the winemaker was, if we are going this far in the vineyard, if we spending all this money, we might as well practice biodynamics in the winery as well."

And now, as I write this, the renowned German, Ernst Loosen, is financing the construction of a fabulous new winery for Jay Somers. It is named not for the German's world-famous label, Dr. Loosen, but for the wild-haired guitar player's virtually unknown label, J. Christopher. It is on a forty-acre property in Newberg with thirteen different soil types. And they are doing this Jay Somers's way. They are farming biodynamically, despite the fact that Loosen doesn't do so at home. They are making the wine according to the principles of biodynamics. When they were clearing old oak trees to plant the vineyard, they found a very old gnarled tree that housed a very old, very large beehive. On Somers's instructions, the tree was uprooted and replanted. The bees remain.

For all these years, the Oregon vigneron Jay Somers has been audaciously avoiding the hype. He has not been playing the game. He has been doing his own thing, in his own obscure corner of the United States, in his own way. He has shown a reckless disregard for the bottom line. But fortune has found him anyway.

He has done it the zen way: He did not aim at the target. The target came to him.

Like many Oregon winegrowers, Somers is an ardent admirer of former Governor Tom McCall, who once famously taunted the rest of the country, "Come visit us again and again, but for heaven's sake, don't come here to live."

Jay Somers—brilliant, stubborn, borderline certifiable at times—is your typical Oregon winemaker.

CHAPTER EIGHT

Big Biodynamics

> Here is a parable of the garden which the righteous are
> promised: In it are rivers of wine, a joy to those who drink.
> —*The Koran*, 47:15

You might recall the James Bond film franchise back in its kookier days, when there was a beloved recurring character called "Q." A mad scientist-cum-engineer extraordinaire, Q developed spying devices that allowed Bond to extricate himself from the hairiest situations with the touch of a button. Thanks to Q's inventions, the old Bond—unlike today's brawny, brooding Daniel Craig—was Inspector Gadget, plus panache. Q's brilliance was in his ability to repurpose everyday objects such as pens, cameras, and cars. He had the vision to see how these items might serve a higher goal, and the know-how to transform them.

On most farms, there's no place for a well-groomed British inventor in a crisp lab coat. But there is most certainly a Q figure at Montinore Estate, the second-largest biodynamic vineyard in the state of Oregon (at 250 acres of vines, it's slightly smaller than Momtazi Vineyard, but unlike Momtazi, it turns all of its own fruit into estate wine): Don Huggett, a gruff, grizzled mechanic with large, chapped hands.

Huggett grew up poor, learning how to make do with what was available. His father was a mechanic, and very handy, and he taught his son how to tinker. By the time he was sixteen, Don had installed Cadillac running gear in a '57 Chevy pickup, creating with his own hands the smoothest and yet most macho ride a teenage boy could ever hope for. Before coming to Oregon, Don owned transmission shops in Fort Collins, Colorado, and Laramie, Wyoming.

When he landed at Montinore Estate in Forest Grove on July 5, 1989, Don Huggett settled in for a quiet career of maintaining farm equipment. But then, in 2005, things changed. A man named Rudy Marchesi bought Montinore and began to convert the estate to biodynamics. And that's when Huggett, an ordinary mechanic, transformed himself into an extraordinary inventor.

Just call him Don Q.

It all started with the tractor attachment for Block Seven, the small triangular section of pinot noir vines planted closest to the winery. With no herbicides allowed on a biodynamic farm, weeds are tilled mechanically at Montinore. But Block Seven is planted like an old vineyard in Burgundy, with low-trellised vines in tightly spaced rows that stand only five feet apart. No standard tractor cultivator attachment fits between five-foot rows. So Don Huggett fitted the bottom of a thirty-gallon drum with a motorbike shock suspension, attaching a lawn-mower deck bearing and dethatcher springs as cutter tines. The result is what he calls "my little turny thing": a hydraulically powered weed eater mounted on a three-point hitch. A gearhead might call this thing a "Frankenrig"—a bunch of junk welded together—but to the staff at Montinore, it represents hours of saved labor.

Another Huggett invention, the "Frodge," is an improbable fusion of the Big Three: a 1975 Dodge motor home, a 1979 Ford van, and the bumper of 1989 Chevy pickup. Equipped with shower, hand-washing sink, eye-washing station, trash and recycling bins, and a covered wagon, it provides escape from the elements plus a few extras for safety and comfort for laborers as they're out working the remote corners of the large estate.

Then there's the "Chaos Vortex," possibly Huggett's greatest achievement. With 250 acres under vine at Montinore, hand-stirring the biodynamic preps was *not* an option. But Montinore operates on a tight budget, so paying thousands of dollars for a custom-built stirring machine or sculpture-like flow form wasn't an option, either. It was up to Don Huggett to invent a machine to perform the laborious task. But there was a large stumbling block: in biodynamics, electricity is not considered to be supportive of the life process.

Huggett didn't ask why the hell he couldn't just set up a straightforward KitchenAid mixer-type device; or, frankly, why this bizarre stirring ritual had to be practiced at all. "Although I don't agree with everything he does," Huggett says circumspectly about his boss, "I just do what he tells me to do. Because he writes the paycheck."

So Huggett shrugged, thought about it, and devised an elegant solution to what appeared to be a ridiculous problem: he designed a *triple-capacity* stirring machine that is far enough removed from its engine that the electrical currents couldn't possibly disrupt the water's dynamic mojo.

Huggett set up three food-grade plastic hundred-gallon water tanks and fitted each with a copper stirring arm attached to a lawn mower gang frame. Then, off to the side and about three feet back from the bins, he installed the four-cylinder motor and frame of a 1989 Ford Motor Company parking-lot sweeper and hooked that up to a hydraulic pump. The hydraulic pump keeps the motor's electrical impulses away from the bins while powering the stirring arms. *Voilà.* Like an observant Jew who pre-programs his lights to turn off at the end of the evening of Shabbat, the machine harnesses electrical energy without coming into visible contact with it.

When I visit Huggett at his shop at Montinore, he is putting the finishing touches on his latest creation: a compost spreader built on the chassis of a 1970 four-wheel-drive Ford pickup. It's got a hydraulic pump mounted on the front crankshaft and two exhaust pipes pointing up out of the front like giant antennae. The back is a massive steel hopper fitted with a conveyer belt and a loveseat-sized red roller brush that disperses manure. Looking at it, I half expect Mad Max to hop on it and drive off into the sunset. But this soldered-together piece of machinery is no joke: unlike the usual slow-going ten-to-twelve-foot-wide tractor-drawn spreaders, this slender custom beauty rumbles easily up and down the property's 220 miles of vine rows, shooting out compost at a healthy clip, then chugging back to the barn to reload in the same amount of time it would take a tractor to turn a single corner.

As Huggett describes his devices, it's clear that he's quite proud of them, if a bit perplexed as to why in the world he has to build them. Although he initially saw it as "witchcraft stuff," his attitude toward biodynamics is now one of good-natured exasperation. He kvetches about the dirty compost teas that the boss has him send through his nice clean sprayers. He groans about the time he had to come up with a "hand post-pounder-slammer-mortar-pestle thing" to crush the quartz sourced from nearby Gales Creek so that the boss could make his own 501 preparation.

Where would biodynamic agriculture be without people like Don Huggett? It would be relegated to small-scale farms staffed by rabid anthroposophists or to the overfunded empires of limousine-liberal environmentalist zealots. But people like Don, who accept it and are willing to work with it—albeit with a raised eyebrow—make it happen on a large yet modestly funded scale, whether they believe in the hocus-pocus part or not.

At Bonterra Vineyards in Mendocino County, California, production is three hundred thousand cases per year—nearly ten times Montinore Estate's. Still, Bonterra farms three of its own properties biodynamically: the McNab, Butler, and Blue Heron ranches, certified since 1996, 2001, and 2009, respectively, and totaling 284 acres. The remaining vineyards Bonterra buys fruit from are certified organic.

When I ask Bob Blue, Bonterra's longtime winemaker, how he's been able to pull off biodynamics and organics on such a large scale, he tells me about the Don Huggetts of his world. They're a support group of people who don't, on the face of it, have anything to do with making wine, or biodynamics. But just like the animals that make up the ecosystem of the ideal biodynamic farm, these are the characters around the edges of the narrative that keep it moving forward.

They're folks like the local 4-H students who come and care for Bonterra's two herds of sheep, which, in turn, aerate the vineyard soil with their hooves and fertilize it with their manure as they feast on the weeds that grow along the vine rows. Or like Thomas Brower of Mendocino Lavender Company, who harvests the insect-attracting lavender fields every summer, then sells lavender essential oils at Mendocino County farmers markets.

Or like Julie Jedlicka, the UC Santa Cruz doctoral student who built two hundred and thirty songbird nest boxes out of local untreated redwood, with holes bored just large enough to attract beneficial birds like bluebirds and swallows but too small to house pesky woodpeckers and starlings. Thanks to Julie, migratory songbirds move into the vineyard birdhouses in the spring, feast on the same insects that threaten the health of the grapevines, and move out by mid-summer, before there's any ripe fruit to snack on.

Having a few of these enablers around—taking care of such seemingly tangential things as sheep, birds, and lavender—makes it possible to farm biodynamically in a big way. And if you can pull this off, you can puncture one of the biggest stereotypes about biodynamic wines: that they're prohibitively expensive.

Montinore can't afford to have gleaming machinery custom-built in a French factory, so Don Huggett rigs something up instead. Bonterra keeps its prices in the supermarket range by building mutually beneficial relationships with students and entrepreneurs—at zero cost to the winery. And small businesses are sprouting up to accommodate the unusual needs of organic and biodynamic vinetenders. An Oregon sheep rancher, Cody Wood, trucks hundreds of his "Green Grazers" to vineyards such as Montinore while the vines are dormant. Section by section, the sheep methodically graze on the cover crops and weeds competing with the vines for nutrients, leaving the tractor lanes clear and naturally fertilizing the soil as they go.

Whether you call it a sign of a new, green economy or a return to the barter system, the result of these relationships is ridiculously affordable biodynamic wine. Montinore Estate makes a killer riesling that sells for less than $10 per bottle; it's one of five tasty white wines they sell for less than $15. Bonterra offers three whites and a rosé for $13 and three reds for $15. And most of the production of Pacific Rim, a Washington-and-Oregon-based partly biodynamic winery specializing in aromatic white wines, sells for just $11 a bottle.

"One reason that biodynamics has caught on in the wine industry, and practically nowhere else, is that wine is perhaps the agricultural product with the largest sales markup," write Douglass Smith and Jesús Barquín in the journal *Skeptical Inquirer*, citing "the onerous biodynamic overhead of labor." The authors go on to question the point of charging "upwards of $50 or $100 for a bottle of what is, in essence, fermented grape juice." The authors should check out the offerings of Montinore, Bonterra, and Pacific Rim, brands that are putting biodynamic grapes in reach of the average grocery-store shopper who isn't going to spend more than $15 on a bottle of wine. And most notably, these large BD producers don't even advertise the fact that they are biodynamic. You have to dig through their Web sites and marketing materials to find that magic word. Neither Bonterra nor Pacific Rim even have tasting rooms where they could educate visitors about what they're trying to do.

So why are they even doing it?

In the case of Montinore, the answer is—contrary to prevailing wisdom—to save money.

Rudy Marchesi's roots are in the Italian pinot noir-producing region of Lombardy, where his ancestors were subsistence farmers. Growing up, he considered his Italian grandparents' place in the Bronx to be a little patch of heaven: their spacious lot grew enough fruit trees, grapevines, and vegetables to keep the family fed for most of the year.

Marchesi had always kept his own garden as a kid and never lost his familial connection to the land, even as an adult. While carving out a successful career for himself as vice president of brand development for Allied Beverage Group in New Jersey, he still had the itch to farm. So, in 1982, he and a friend planted vines and launched a winery on the Delaware River. Alba Vineyard is still a successful operation, but as much as he enjoyed managing the vineyard and making wine, Marchesi found it frustrating that he couldn't farm naturally, the way his grandparents had. Milford, New Jersey, "was really the wrong place to grow grapes," he says now in retrospect. "The disease pressure was so high, with the high humidity and the warm nights in July and August, it would have been an effort that I'm not sure I would have been successful at."

Meanwhile, an Oregon winery that had hired him to do some marketing consulting was frustrating him for a different reason: this pristine vineyard planted in an area with optimal weather conditions for grape growing was producing mediocre grapes and uninteresting wines. Why? Intrigued, Marchesi sold out to his Alba Vineyards partner in 2001 and moved to Oregon to oversee winemaking operations at Montinore.

In 1905 John Forbis, a corporate attorney for Montana's Anaconda Copper Company, had built a magnificent ten-thousand-square-foot mansion on the property, naming it Montinore for "Montana in Oregon." Leo and Jane "Bobsy" Graham acquired the 361-acre estate in 1965 and renamed it "Dilley Farm," adding additional acreage; by 1972, the property encompassed 588 acres. In 1982, they hired winemaker Jeffrey Lamy, who directed the planting of French grape varietals in the most vineyard-worthy sections of the farm.

The estate was magnificent to look at. But after an initial string of critically acclaimed vintages, Lamy left in 1992 and the quality of the wines began to diminish. By 2005, Montinore was known as a beautiful place for a wedding where you might not want to serve the wines.

Upon his arrival at Montinore Vineyards, Rudy Marchesi immediately noticed two things. First, the site was perfect: gently sloping and protected from weather by Bald Peak and the Chehalem Mountains, it centered around a fifteen-acre reservoir that attracted enough beneficial birds to make pesticides unnecessary. Second, the vineyard was in terrible shape: The vines looked pale and limp. The devastating root louse, phylloxera, was eating its way through the older sections. The ground was rock-hard and riddled with gopher holes; water ran right off of the earth instead of soaking in. "We had serious compaction, very anaerobic conditions in the soil," recalls Marchesi. "There were just a lot of things in vineyard management and plant management that needed to be changed."

To the surprise of his employers, Marchesi immediately halted the use of herbicides on the site and switched to organic farming techniques. But this didn't seem like a bold enough step to him. "Organics is the same logic as what we now call conventional agriculture except that it uses softer materials," he says. "I knew there was a different way to approach all of this ... I thought, 'Do you take an aspirin or do you cure the disease?'"

Remembering his grandparents' garden and the delicious biodynamically produced French wines he had tasted during his time at Allied Beverage, Marchesi wondered if he could introduce the same old ways of farming at Montinore. "I had worked in conditions where powdery mildew was the easy one. We had downy mildew, we had black rot, we had phomopsis, we had crown gall. We had serious problems," Marchesi recalls of his New Jersey vine-tending days. "So to come to this area where the biggest deal was powdery mildew and some botrytis, I thought, 'This is heaven. I can handle this. And we can handle this in a much more gentle way than the chemical companies are telling us.'"

In 2003, he enrolled in a one-year training course in biodynamic farming and gardening at The Pfeiffer Center in Chestnut Ridge, New York (incidentally, just five miles from his childhood home), and simultaneously began implementing what he was learning at Montinore. The results he saw were promising, but Marchesi didn't feel that he had the authority to turn the entire estate's farming system on its head. Then an opportunity presented itself. The Grahams, knowing that their children did not want to continue in the wine business, put the property up for sale. Their $5 million asking price was at the time unheard-of in Oregon. But considering the size of the property and the capacity of the winery, Marchesi realized that his bosses were offering a bargain. He shrewdly negotiated to purchase the grapevine blocks and the winery, leaving the Grahams with the palatial manor house and additional acreage.

Thus—unlike the typical new winery owner who sinks millions of dollars into luxurious new facilities and virgin vine plantings years before ever seeing profits on his investment—Marchesi instantly became the proprietor of a fully functional operation, including a well-equipped winery and 230 acres of established vines (he soon planted twenty more). Now it was time to get to work.

Marchesi didn't have any money to spare, which was fine, because he didn't need to invest in infrastructure, or in purchasing fruit from independent vineyards. All he and winemaker John Lundy had to do was make their mediocre estate wines taste better. And they already knew how they were going to go about it: in 2006, they converted the property to biodynamic viticulture.

Today, the grape leaves are bright green, the vines perky and alert. The soil is soft and fragrant thanks to a regime of tillage and cover crops. And the rodents have moved on. "My theory is that we were so compacted before that the gophers were trying to aerate the place for us," Marchesi says, only half joking. Most important, the wine quality has improved dramatically. And, according to Marchesi, it is costing less to produce.

For example: When phylloxera hits a patch of older vines, it's standard procedure to rip them out and replant on different rootstock. But this is a time-consuming and costly process, especially when you consider that it takes three years or so for a newly planted vine to produce fruit. So Marchesi hasn't replaced any of Montinore's old louse-ridden vineyard sections. "What the phylloxerae do is create an open wound where pathogenic fungi can get in," Marchesi explains. "We are now through biodynamics introducing beneficial mycorrhizal fungi that are crowding out the pathogenic fungi. They are making the soil more aerobic. Pathenogenic fungi like soils that are anaerobic." Looking out at the vineyard, there is still one scraggly section affected by phylloxera; the other formerly ailing sections now look as green and vigorous as unaffected plantings on rootstock.*

While neighboring Willamette Valley vineyards spend an average of $4,500 an acre on farming, Montinore spends just $3,200 an acre. Ask vineyard manager Efren Rosales, who has worked on the estate for three decades, what's changed since 2005 and his answer is simple: he has reduced his crew by seven farmhands. "Because the vineyard is a lot easier to take care of now," he explains. "Before, there was more stuff in the vineyard that we had to deal with. Now we have control and it is a lot easier."

The savings are also cropping up in the winery, according to winemaker John Lundy. Grapes are coming in uniformly ripe these days, without botrytis or mildew. The harsh, astringent tannins Lundy once struggled to mask are now naturally soft and gentle on the palate, the fruit flavors fresh, the acidity bright. Lundy doesn't waste time problem solving and purchasing oeno-chemical products to correct faults the way he used to, because his grapes now come in clean. "We're no longer having to guess what the finished wine is going to be like, because we taste it out in the vineyard," the winemaker says. "In the winery, it's now really about minimalist handling, about just letting the farm show through."

In short, Montinore is achieving the unthinkable: its wines are biodynamically farmed and made "naturally" in the winery, like so many of the European wines that garner top scores from the critics and sell for more than $50 per bottle. And yet, it sits modestly on the shelves of American grocery stores, most of it priced at less than $20. When some *MIX* magazine colleagues and I recently blind-tasted through nearly thirty Oregon pinot noirs in the $20 price range, we chose the 2008 Montinore as our favorite. By the time the article hit the magazine rack, the price had been reduced to $16.

Of course, if it were a simple question of economics, many more large wineries would farm biodynamically. Montinore may have been a fluke, an estate uniquely poised to embrace biodynamics and cut

---

* The vines in the Willamette Valley that are attacked by phylloxera are the older, self-rooted plantings. Since 1990, when phylloxera was discovered in Oregon, *Vitis vinifera* (wine grapes) have been grafted onto disease-resistant rootstocks prior to planting.

costs all in one fell swoop. Vintners on other sites in Oregon report that implementing biodynamics is an expensive proposition. And remember: only Montinore has Don "Q" Huggett on staff.

But there are other large-scale winemaking operations quietly adopting biodynamics for their own reasons. Consider Pacific Rim, the Portland- and southeastern Washington-based winery that produces one hundred and ten thousand cases annually—more than three times the volume at Montinore. Pacific Rim is an offshoot of Bonny Doon Vineyard, the Santa Cruz-based oeno-phenomenon led by quirky visionary and biodynamic believer Randall Grahm. The brand centers on riesling—incidentally, the wine James Bond savors in Ian Fleming's 1959 novel, *Goldfinger*—and other delicate white wines that pair well with Asian cuisines; its predominant price point is $10 per bottle.

Its corporate offices are in a LEED-certified building on hipster-heavy East Burnside Street in Portland; its $5.7-million winery is in Washington's Tri-Cities area near the acclaimed Wallula Vineyard, which custom-planted 140 biodynamically managed acres of riesling exclusively for Pacific Rim in 2005, supplying the winery with more than a third of its juice. Pacific Rim's block of Wallula is one of only two biodynamically certified vineyards in Washington (as of this writing), thanks to the fact that the riesling vines have been trained onto unusually high sixty-six-inch cordons, so sheep can chow on weeds and prune suckers any time of year—even when tempting ripe fruit hangs from the vines, it's just out of reach. It is biodynamic thanks to an $80,000 custom copper stirring machine imported from France. It is biodynamic thanks to a speedy specialized tractor (price tag: more than $375,000) that can spray six rows of vines at once and doubles as a mechanical harvester.

(Yes, you read right: If you want a nationally available $10 riesling made from biodynamic grapes, you'd better accept mechanical harvesting.)

Winemaker and general manager Nicolas Quillé is proud of all the expensive, gleaming equipment that makes his winery so efficient and, incidentally, look like it could be Dr. No's nuclear-weapons facility: antiseptic and brightly lit, it's all stainless steel and blinding white. Quillé embraces the refrigeration that allows him to make crisp riesling in the early-autumn heat of eastern Washington. He proudly points out a centrifuge (approximately $120,000) that eliminates filtration waste and an electrodialysis machine (a cool $250,000) that uses a fifth of the energy of traditional cold-stabilization processes.

Quillé, a Burgundian by birth, has a goofy sense of humor; one can see why the famously irreverent Randall Grahm enjoys his company. But he is also a pragmatist with a sharp brain and a forthright manner. When I ask him about all the technology Pacific Rim employs, he shrugs. "We're using clean, energy-efficient equipment to make better wine," he says.

But Quillé doesn't always make use of his technology. When Pacific Rim releases a single-vineyard, certified biodynamic bottling from Wallula Vineyard fruit, it's composed of the select group of grapes that the winemaker deems so perfect that they need no technical assistance, other than refrigeration. But it's priced at $32. If you want a luddite wine, you have to pay for it. "There is something interesting about making a wine with almost no intervention. It may not be better, but it is interesting," says Quillé. "There are a lot of tricks to the trade and a certified biodynamic wine doesn't make use of any winemaking tricks. Every once in a while, you should be able to try something that is trick-free." Most of us can't afford to drink a trick-free $32 bottle of riesling on a weeknight. That's why Quillé's goal is to produce boatloads of delicious, affordable riesling, from grapes farmed as naturally as possible. If that means mechanically harvesting grapes after the vines have been pruned by the mouths of sheep, so be it. Life is full of ironies.

"At the end of the day, biodynamics are not what will make a wine great or not great," Quillé says resolutely. "It is more a question of philosophically what you want to do with your life. If you spend your days mining oil, are you going to feel good about what you've done? I wouldn't. If we can farm that land economically and not dump tons of chemicals on it, why not do that? Philosophically it sounds very good to me."

CHAPTER NINE

The Glam Factor

"Pinot noir vines, if you don't trellis them, will fall on the ground. These vines are really terrestrial. They're linked to the earth, but we are helping them to reach to the skies and their etheric forces. And at their last gasp for the year, they offer up this beautiful fruit. And each grape is created to encapsulate this little tiny seed encoded with the vine's entire genetic DNA.

"Look at a vine: it's green and brown—earth colors, not these glorious sunset colors like you see in flowers. But these beautifully colored grapes, they're almost like the perfect transition between the terrestrial and the heavenly. We're trying to capture that transition or that joining point. As winemakers, we're trying to encapsulate that year of sunlight and of water—of both forces working on this one plant. And then transmute them through this alchemical process of fermentation and put them in this bottle. ...

"No other consumable thing has that same distillation of heaven and earth."

—Sam Tannahill, director of viticulture and winemaking for the wineries A to Z, Rex Hill, and Francis-Tannahill

Oh, to be Sting and Trudie Styler! To be a rock star and an actress/producer/yogi! To have a gaggle of children and a private plane and a staff of domestic assistants and spiritual advisors and seven homes! And for one of those homes to be a nearly one-thousand-acre estate in Tuscany, a working farm under the watchful eye of superstar biodynamic consultant Alan York, producing extra-virgin olive oil, chestnut and acacia honey, jams, fruit, vegetables, and salami and—best of all—biodynamic Chianti!

Oh, to be Danielle Andrus Montalieu, the platinum-blonde daughter of the Olympic skier who founded Napa Valley's storied Pine Ridge winery and the Willamette Valley's luxury label, Archery Summit! Oh, to be sitting at a patio table by her pool, overlooking her picturesque and biodynamically farmed Domaine Danielle Laurent vineyard and Soléna winery and Grand Cru Estates!

And oh, to be Paul and Kendall Bergström de Lancellotti! He, a tan former globe-trotting surfer with a name worthy of an Arthurian knight, and she, a blonde former pro snowboarder and member of the renowned Bergström winemaking family! They, the founding partners of the Inn at Red Hills and Farm to Fork restaurant in Dundee! They, who live on a biodynamic gentleman's farm and produce $65 bottles of de Lancellotti biodynamic pinot noir that the critics go gaga for!

Oh, to be Mike Etzel, brother-in-law and business partner of the world's most powerful wine critic, the great Robert Parker! And to be able to charge $90 for a bottle of one's outrageously delicious and aptly titled biodynamically farmed Beaux Frères wines!

And, oh, to be James Frey, with his Superman looks and sleek new winery outfitted with a personal art gallery! And to own two vineyards brimming with some of the best dry-farmed, biodynamic fruit in the Willamette Valley, and to have one's $75 pinots getting 90-plus scores from Parker, straight out of the gate!

Oh, to be blessed with good genes, good fortune, a good work ethic, and good credit, because if you have these four things, perhaps you, too, can make biodynamic wine!

It is tempting to deem the wealthy wineries that have taken a shine to BD as merely a group of beautiful people dabbling in something fashionable. One might harbor momentary visions of Marie Antoinette, costumed as a shepherdess, play-milking docile cows on her fanciful farm, the "Hameau de la Reine," that rustic pseudo-hamlet on the grounds of the Petit Trianon at Versailles—in tune with her own ersatz notion of nature but completely out of touch with the ways of the world.

It is tempting, but it is wrong. Because what unites the high-end biodynamic wineries is not their faddishness, but rather their pursuit of perfection.

Mike Etzel, for example, would laugh if one were to accuse him of being one of the beautiful people. His skin has the leathery quality of someone who started his career on a Maryland dairy farm and continues to spend most of his time outside, on a tractor. There is an impishness to his speech that makes much of what he says sound as though it were said in jest. But tell Etzel that his fellow vignerons consider him a perfectionist and he falls silent. Because he is, as Alan York would put it, anal. And he knows it.

For Etzel, making great wine isn't enough. So he coddles his vines with homeopathic teas and more compost and cowhorns than are really necessary or called for (forty cowhorns for thirty-three acres of vines, for example). "My intention is to make the best possible wine. I believe that biodynamics can do that. I think what makes people like us farm this way is questing, searching for the absolute very best way to produce a product that is true and honest and pure," he says.

Considering that his estate consists of two very small vineyards, Etzel's two compost-processing centers look like the work of an obsessive-compulsive. One consists of rows of schoolbus-sized piles of carbon-based wood compost (i.e., sticks, grape canes, and branches cleared from the estate) to apply to over-vigorous vineyard blocks; in the other, equally massive mounds of nitrogen-rich manure-based compost steam in the afternoon heat, ready to prop up areas weakened by phylloxera.

Etzel's team uses an antique firetruck to spray a thousand gallons of water every two weeks on each of these piles to maintain aerobic decomposition conditions; according to Etzel, the water actually raises the temperature of the pile and encourages fermentation of its contents. "This is like liquid gold to me. I store it like money in the bank," Etzel says, picking up a handful of fine brown compost and letting it fall between his fingers.

Understandably, Etzel shies away from discussion about his "beau frère," the famous wine critic Robert Parker. But he does credit Parker with handing him a copy "a long time ago" of Rudolf Steiner's lectures on agriculture. "There is a common denominator of great wines coming from great vineyards: great vineyards are attentively farmed. And attentively farmed vineyards are organically and biodynamically farmed. So there is a relationship and he is excited about that," Etzel says of his brother-in-law.

When he began practicing biodynamics in 2003, Etzel's Beaux Frères wines had already garnered high scores. They already sold in futures. But the vigneron wanted more. And so, since he can't control the weather, Etzel now controls the birds: barn swallows, attracted by birdhouses mounted on wild cherry trees and pear trees throughout the property, sweep down from the heavens every morning and evening, plucking insects from the earth.

The environmental, foodie, and feminist movements all started among the economically advantaged, but have trickled down to the masses—any modern-day working woman can pick up dinner at Chipotle, for example, where the meat comes from cage-free, vegetarian animals, and 40 percent of the beans are certified organic.

Organic is ubiquitous. You can find Amy's Organic Soups at Wal-Mart and buy an organic cup of Green Mountain Coffee Roasters Joe at McDonald's. Safeway and Albertson's each stock their in-house organic brands—O Organics and Wild Harvest Organics, respectively—on nearly every aisle of every store. And that includes the wine aisle, with widely distributed labels such as Fetzer now dropping the "O" word.

"Organic," in short, is special no longer.

But "biodynamic" still occupies its own rarefied realm. Even if it's affordable, it's a term that hasn't trickled down yet. You really have to scour the aisles of Whole Foods Market to find that precious tin of biodynamic "couture tea sachets" from Zhena's Gypsy Tea. If you're very lucky, you might find the tender, delicate, and shockingly sweet Demeter Certified Biodynamic green grapes from Marian Farms in Fresno in a produce bin.

Stroll through the skincare section at Whole Foods and you might spy tubes of luxe lotions labeled Weleda. This Switzerland-based, Steiner-founded company produces pricy salves (Pomegranate Firming Day Cream: $33 for 1 ounce) used by famous beauties such as Kate Hudson, Julie Delphy, Demi Moore, Alicia Silverstone, Claudia Schiffer, Rihanna, and the designer Tory Burch. Of course, to get the full Weleda experience, it's best to travel to New York, Paris, or Tokyo and luxuriate in the biodynamic treatments at a Weleda Spa. Likewise, *Allure, Elle, InStyle, Lucky, Marie Claire* and *Town & Country* are among the many publications that have gushed over Jurlique, a high-end natural skincare company that grows its ingredients on its own biodynamic farm in the Adelaide Hills of South Australia (Soothing Day Care Lotion: $40 for 1 ounce). Can't find Jurlique? Another high-end skincare brand you might find at Whole Foods Market is Dr. Hauschka (Regenerating Day Cream: $80 for 1.35 ounces); according to *In Style* and Style.com, Jennifer Aniston, Madonna, J. Lo, and Uma Thurman swear by the stuff. It's made from biodynamically grown herbs and flowers, but of course.

Forget about hemp-wearing hippies. Consumers of biodynamic goods are "the same people who are going to farmers markets and looking for locally made goat cheese. They are aware of their surroundings and what they are putting in their bodies," observes Amy Atwood, a California sustainable-wine importer and wholesaler who blogs about organic, biodynamic, and natural wines.

Flipping through a copy of *The New York Times* or *The Wall Street Journal*, you might happen upon an article about Ubuntu, the Napa, California, Michelin-starred restaurant, where the fresh, critically acclaimed cuisine is created using biodynamically gardened produce. It's merely a reflection of what is going on in wine country. "More and more vineyards have dramatically improved their health through the so-called green movement, moving toward quasi-organic farming, out-and-out organic farming, or to the most extreme style, biodynamic farming," writes Robert Parker, guru to wealthy wine collectors. "… There is no question that the result is better fruit and, as a consequence, better wines."

Publications for serious oenophiles—*Food & Wine, Decanter, The Wine Spectator, Wine & Spirits*, and *Wine Enthusiast*—no longer bother to define the term, because their readers are wine insiders. And wine insiders know all about biodynamic viticulture. "Is it an intellectual's wine? Yes, of course it is, since there are very few consumers out there who understand what biodynamics is. And, yes, it is a step up in quality and in price as well, because by its nature it's hard to get to huge economies of scale," says Atwood.

Half a mile east of The Beaux Frères Vineyard is Trisaetum winery and its Ribbon Ridge vineyard. Visitors entering the new thirteen-thousand-square-foot winery are greeted by a massive three-panel oil triptych, approximately fifteen feet wide, depicting the winery's other property, the Coast Range Vineyard. Fitting right in with his suave surroundings is owner James Frey, surprisingly clean shaven and impeccably groomed on a typical workday for winegrowers. This isn't because Frey doesn't get his hands dirty in the winery and vineyard; he does. He is simply a guy who takes the definition of the word "perfectionist" to another level. His winery appeared seemingly out of nowhere in 2005 to scoop up huge scores from the critics for its precocious pinot noirs and rising-star rieslings.

A quick walk through the winery confirms one's suspicions that it's a well-funded venture. There's an impressive *cave*, full of costly French barrels, each printed with the winery's logo. There are the new two-ton oak fermenters in the sixty-five-foot-long tank hall. There's a cold room, where just-harvested fruit is chilled to precisely thirty-five degrees, to allow for a slow and even fermentation.

But it's the touches that a casual visitor might not notice that point to Frey's neat-freak streak. The roof is inverted to collect rainwater, which is cleaned by a reverse-osmosis system, then used to rinse down winery equipment. An AiroCide photocatalytic air purifier powered by UV light keeps the air mold- and microbe-free. And then there's the nine-step—let me repeat that: *nine-step*—sorting system.

Sorting through grape clusters at harvest time is typically a matter of clumsily rubber-gloved fingers and teary-red, allergy-bothered eyes frantically trying to pinpoint and remove desiccated, rotten, under-ripe, or moldy fruit, as well as any stray MOG—material other than grapes—before a conveyor belt swiftly deposits the bunches of fruit into a big, clumsy crusher-destemmer. At Trisaetum, there is no MOG. Because anything that isn't part of a grape cluster—leaves, twigs, weeds, yellow jackets, ladybugs, spiders, and tons of earwigs—is removed, quickly and efficiently, by "the bug sucker," as Frey calls it. This high-powered vacuum gently deposits the vermin, alive and kicking, along with any other stray debris, into a compost bin, which is carted outside and dumped onto a compost pile.

The fruit then passes under an air knife, where it is cleaned of dust and condensation. Only then do human hands perform the first cluster-sort, after which the grapes go up a conveyer belt into a destemmer. Then comes the second, single-berry, sort, which reduces the fruit load down to just a few clean, glistening globes, before it is crushed. "We're very inefficient, but we're trying to be very effective," Frey says.

By the same token, Frey has planted his two vineyards in the Burgundian style, with tightly spaced rows and no irrigation to fall back on in hot years. His trellises are slightly shorter than those of his neighbors, so that each row doesn't cast too much of a shadow on the next. But even this isn't enough for Frey. "You've got to have the right site oriented the right way, at the right elevation, with the right soils. And you've got to have the right plant material and rootstock. Dry farming was important to me, to really stress the vines and force the roots deep. So we dry farm both of our vineyards. And then we were organic, because we believe in minimal intervention: no pesticides and no herbicides. And then, once you've got all those criteria met, then it's like, 'Is there a little bit more that we can do?'"

For now, that's why Frey is dabbling in biodynamics. He's got a stirring machine and he's applying the preps, seeing what the effect will be on his still-young vines. He's aware that some of the aspects of biodynamics can seem ridiculous—"We don't say any prayers. We don't spread any ashes. I haven't started spraying cow urine yet. We have dogs who pee on the grounds. Does that count?"—but determined to stick with the practice if it means that his wines will taste more like ambrosia. "If someone can prove to me that it really works well, we'll do it," he says. "Our goal is to make a premium product, so we need to do whatever it takes."

Is Frey practicing biodynamics for marketing reasons, as critics of high-end BD producers so often charge? When one considers how many extra steps are quietly taken at Trisaetum, it's difficult to level such charges. For example, when I point out to Frey that he could advertise the fact that his "bug sucker" might make Trisaetum the most humane winery in the state, he demurs. "We do these things because we believe a couple of extra steps might help us make better wines," he says. "If it means we could classify ourselves as vegan, that's interesting, but it's not why we do it."

Trisaetum isn't certified. Frey is just dabbling for now, seeing if biodynamic agriculture is the factor that can take his wines over the edge, from great to truly magnificent. At Beaux Frères, Mike Etzel willingly admits that, because he occasionally makes adjustments during vinification*, his winery wouldn't merit certification even though his vineyards would pass with flying colors: "I don't make the wines biodynamically; I just grow the grapes biodynamically."

---

* The regulations for Demeter-certified "biodynamic wine" labeling prohibit, among other things, acidity and sugar adjustments, micro-oxygenation, cold stabilization, and the addition of cultured yeasts. Although the rules are less stringent for wines marked "made with biodynamic grapes" (often the label is merely stamped with a small Demeter logo in place of these words), commercial yeasts are still prohibited except in extreme circumstances, and many other winemaking activities are regulated.

It's possible that Etzel hasn't rushed to be certified, also, because he is a disciple of the French biodynamic consultant Philippe Armenier, who, while not exactly against certification, also likes to say that "biodynamics stop at the cellar door."

Tall and broad, with deep brown eyes under a thoughtful brow and a shock of white hair, Armenier comes from a family that was tending vines in Châteauneuf-du-Papes as far back as 1344; his sisters continue to run Domaine de Marcoux, the Armenier family estate.

After converting Marcoux to biodynamic viticulture in 1990, Armenier followed his wanderlust throughout France and Europe and around the world, studying biodynamic agriculture in Australia and Israel with his wife and children in tow. He ended, finally, in Santa Rosa, California. From this home base, he travels constantly between his thirty-some vineyard clients located up and down the West Coast.

Armenier is a Steinerian fundamentalist, firmly steeped in the anthroposophic lifestyle (he and his wife were part of a group that founded a Waldorf school in France). But his interpretation of BD scripture is expansive. In fact, it's so expansive that he's willing to think of the United States as one giant farm so that his clients can order their preps from the Josephine Porter Institute in Virginia without feeling guilt about not producing them on site. Likewise, Armenier feels that the Demeter certification rules restricting the winemaker's freedom within the cellar walls are too stringent. "For me, there is no such thing as biodynamic wine; there is only wine made from biodynamic grapes," he says in exasperation. "In France, my vineyard was Demeter certified, and I could put that on my front label. Here if the winery is not certified, you cannot write it."* As far as the Frenchman is concerned, after the juice has fermented into alcohol, it is, in effect, no longer living—"If you want to keep something from rotting, you put it in a glass jar in alcohol and it will stay the same forever"—so whether or not an oenologist wants to add some enzymes, sugar, acid, or non-native yeast to his wine is immaterial to its biodynamic-ness. "You don't spray the preparations in the winery," he points out.

---

* Even a wine labeled "made with Biodynamic grapes" must be made in a Demeter-certified winery.

Thus, the followers of Armenier quietly observe the dictums of Steiner in their vineyards, but may or may not follow the dictums of Demeter in the winery. And many don't bother to get certified. They're already paying Armenier a per-acre fee (while Armenier is coy about how much he charges, one winery reports that they have paid him between $6,000 and $10,000 annually, depending on the number of acres he has overseen*); why should they dole out even more dollars to Demeter USA?

And so, among Armenier's clients are high-end labels not generally known to be practicing biodynamics. Such as Archery Summit, producer of a sought-after estate pinot noir that sells for a cool $150 per bottle. For three years (2004-2007), Armenier helped vineyard manager Leigh Bartholomew to establish a biodynamic regime on the winery's Renegade Ridge Vineyard, a property that was living up to its name with wild overgrowth that couldn't be reined in and that produced disappointing fruit. After the implementation of BD, says Bartholomew, "[i]t started doing the things we wanted it to do: the flavor profiles seemed brighter and more balanced. It's a cliché to say it felt more 'alive,' but there was definitely something more interesting about the fruit afterwards." Later, "We had a soil scientist come down from Canada. He was looking at soils and we didn't mention that we were doing biodynamics at all. We went to Renegade Ridge and he said, 'Wow, this soil is so different: it's so alive, it's so easy to dig,'" Bartholomew recalls. "Now it's the tasting room staff's favorite wine, which usually that means it's one of our best wines." Price tag: $85 per bottle. And the soil scientist told Bartholomew that he'd had such a good time that he wasn't going to bill her.

Vineyard saved, profitable wines made. And the investment in Armenier had a trickle-down effect, as well: Bartholomew's husband, Patrick Reuter, tagged along on as many of Armenier's visits as he could—"Until he realized I wasn't going to be a client, and then kicked me out," Reuter chuckles—soaking up the teachings of the master. Today, in their spare time, Bartholomew and Reuter practice biodynamic viticulture on their own Demeter-certified Three Sleeps Vineyard in the Columbia Gorge, an improbable piece of proof that there may be something to the theory of trickle-down economics:

---

* That said, Armenier is such a passionate believer in BD that he doesn't charge his clients when they are unable to pay.

Archery Summit can charge ungodly sums for its wine, and so it can afford to hire Armenier. Armenier doles out his wisdom, and Reuter then puts it to use in his own project.

The jewel in Archery Summit's crown is its Arcus Estate vineyard in the Dundee Hills. The northern neighbor of Arcus is Winderlea, a self-proclaimed "luxury boutique winery" on the esteemed piece of land formerly known as Goldschmidt Vineyard and originally planted in 1974. (Another biodynamically farmed vineyard, Holloran's Le Pavillon, is just across the street.)

When I visit their sleek contemporary winery, Winderlea's owners, Donna Morris and Bill Sweat, are dressed in a casual-chic style that screams "early retirement": Bill is in a fitted gray T-shirt and dark jeans while Donna wears a soft chocolate cowl-neck shift over black leggings, her hair artfully cropped. When I arrive, the couple is meeting with their tall, silver-haired winemaker, Robert Brittan, who left Stags' Leap Winery in the Napa Valley to come north and make pinot noir. Behind them, visible through a wall of windows, is a dramatic backdrop: rolling vineyards, forest, and the valley floor below. I feel like I've walked into an ad for Fidelity mutual funds.

Which Winderlea is, sort of. Because Fidelity Investments was Sweat's and Morris's employer, until they decided to leave the East Coast rat race behind and make wine in Oregon. "Donna and I were given the opportunity in 1997 to work in Tokyo," Sweat recalls. "We got all our ducks in a row and created wills. And then we decided to create a goal for ourselves to leave the corporate world in ten years. When we got back after three years in Japan, we started running through different ideas, some of which would have made our financial advisers a lot happier than what we chose. But we kept coming back to wineries. We collected wines and did wine travels and stood on other peoples' vineyard decks and said, 'We should do this.'"

By 2006, Bill and Donna were ready to make the leap. A February 2007 *New York Times* article shows the couple in their new cellar, thoughtfully tasting chardonnay out of the barrel. "Donna Morris and Bill Sweat were in finance," the caption reads. "Now they make wine."

The couple has been, wisely, sinking money into all the things that limousine-liberal wine aficionados approve of.* Such as a winery and tasting room that's solar powered and illuminated by daylight, and includes an electric vehicle-recharging station in the parking lot. They've signed on to the state's Carbon Neutral Challenge, and they're major supporters of ¡Salud!, a glamorous black-tie wine auction and multi-day fundraiser that provides healthcare services to uninsured agricultural laborers.

"There was never a question that we would be as sustainable as we could be," says Morris. Which is why, she adds, they've begun farming biodynamically. "To me," says Morris, "biodynamic viticulture it is all part of our ethic of sustainability. We want to be as soft on the property as we possibly can, so that we can create a vineyard that is more self-sustaining. The more we read about biodynamic practices, the more they seem to hold to the philosophy of trying to create a self-sustaining environment." And so, the Winderlea Web site includes a page entitled "Biodynamic® Consultant," featuring a photo and glowing bio of the sage from Châteauneuf-du-Papes: Philippe Armenier.

Another couple of apparently glamorous Armenier clients are Danielle Andrus Montalieu and her dashing French husband, Laurent. After successfully starting up and selling the renowned Pine Ridge winery in the Napa Valley, Danielle's father, former US ski team member Gary Andrus,** founded the Archery Summit winery in 1993, with the then-twenty-five-year-old Danielle working by his side; when Gary sold his shares in 2001,*** Danielle stayed on during the transition period. Today, the Montalieus own—and biodynamically farm—their own eighty-acre**** Domaine Danielle Laurent and co-own Hyland Vineyard, a nearly two-hundred-acre property southwest of McMinnville that includes one hundred acres of vineyard planted in 1971. The couple produce their own Soléna wines, named after their daughter, who, in turn, was named after the sun and the moon.

---

* Lest she be accused of sarcasm, the author feels compelled to note that she shops weekly at Whole Foods Market, purchases goat cheeses at farmers markets, and greatly enjoys all the wines featured in this chapter.
** Andrus, whose full name was Robert Gary Andrus, died in 2009.
*** Both Pine Ridge and Archery Summit are now owned by the holding company Leucadia National Corporation.
**** Twenty-two acres under were under vine as I was researching this book.

Laurent also manages Doe Ridge Vineyard in the Yamhill-Carlton District for the high-end Four Graces winery; he's farming half the property (twenty acres) biodynamically and comparing the results with an equal-sized, sustainably farmed parcel. Additionally, he's the founder and managing partner of NW Wine Co. in McMinnville, which custom-produces close to one hundred thousand cases of wine each year for twenty-five different clients, and the adjoining NW Wine Bar. Oh, and the couple is also selling luxury home sites on nine parcels of Hyland Vineyard. "It seems like every year we've been married, we've started a new project," Danielle says with a laugh.

The final piece of their entrepreneurial puzzle is Grand Cru Estates, which offers private memberships for between $6,000 and $20,000 per year to the lucky few who can afford to have Laurent Montalieu and his fellow winemaker, the critics' darling Tony Rynders, create custom barrels of pinot noir for them from a selection of seventeen different vineyard sites. The handsome Grand Cru winery,* adjacent to the prestigious Shea Vineyard, sits on a one-hundred-thirty-acre parcel that includes the Domaine Danielle Laurent vineyard and the Montalieus' well-appointed home. The winery boasts a professional kitchen, a massive stone fireplace, granite countertops, soaring, open-beamed ceilings, and the now-*de rigeur* green touches of solar panels and rainwater collection.

Ask the Montalieus why they are dabbling in BD when they're already so busy with other ventures and they'll tell you about Laurent's intense conversations with Jimi Brooks prior to his tragic death, and about the stunning wines from France that come from biodynamic vineyards. They'll tell you about the joyful moments they've spent quietly at home, as a family, working. Stirring the preps. Feeding their goats. "Soléna helps me apply the sprays; she rides with me the whole time. It's really a connection: father-daughter, mother-daughter, vineyard, whole earth," says Laurent.

"Our purpose is health of our land, the health of our vines, and giving something back to our daughter," adds Danielle, her lustrous blonde hair gleaming in the morning sunlight. "And doing something that is sustainable for the valley. This is a crop that lasts forty or fifty years before it has to be replanted."

---

* Grand Cru and Soléna share the same building.

"You know Danielle Montalieu? If you saw her out on the town or at a ¡Salud! auction, looking fabulous, she is not the person you think of piling manure or mixing the preps," observes David Millmann, managing director of Domaine Drouhin Oregon (and, it should be noted, no fashion slouch himself). "Then she talks about her and Soléna mixing together, and I think that's just great. I know it's good for the vineyard, but look at the impact it has on her family. That's the cool thing: it's a family event. It's like the way that yoga seeps into certain families: it's just healthy."

It also must be healthy for a high-end winery to have a biodynamic aspect to boast about. Because a Grand Cru membership includes hang-out privileges at the estate, with its culinary program headed up by James Beard Award-winning chef Philippe Boulot, its personal wine-country concierge, and—but of course—its biodynamic vegetable gardens.

Plebes can have access to biodynamic vegetables, as well, if they stop in to dine at Farm to Fork Gourmet Café & Market in Dundee, where no membership is required. The restaurant-cum-wine-and-gourmet-shop was the brainchild of Kendall Bergström de Lancellotti. Her husband, Paul, is busy prepping the quarter-acre biodynamic vegetable garden on their nearby biodynamic vineyard estate to supply produce to the eatery when I meet with him.

Considering his background, it's difficult to imagine Paul de Lancellotti farming any other way: born and raised in Newport Beach, California, he spent his early career on his own present-day Arthurian quest, traveling around the world, surfing and representing the Quiksilver brand of surf gear. So when he talks about astrology, it sounds … natural and groovy. "I'm a Cancer," he says. "I'm somebody who loves the water. For me, surfing has always been a religious experience. Being out in the ocean and understanding that water is 80 percent of this planet. Because of my connection to the ocean, I've always felt a connection with the earth. In the leaf, we have the element of water. In the fruit, we have the element of heat, which I love. We have these elements in biodynamics which are in all of us."

Cue the soundtrack: the sound of rolling waves, maybe some ukulele riffs … "The more we can start aligning these forces, the sooner we can start healing the earth from the inside out," de

Lancellotti continues. "Steiner knew that global warming was coming. He knew that bee populations would be wiped out. I believe there are people out there who have a high level of energy and are connected. I have a friend who is a healer. People have things within them that sometimes get tapped."

De Lancellotti, whose favorite word is "passionate," is a walking narrative trope: like the Arthurian knight he really ought to be with a name like that, he's a seeker and adventurer, with a charming touch of innocence. But there is nothing naïve about the de Lancellotti Family Vineyards label, whose small output ensures that it will be a collector's item for years to come. After eighteen to twenty months of barrel aging, the wine is bottled on an astrologically auspicious day, with classical music on the winery stereo. Each cork is sealed in blood-red wax, and those precious $65 bottles are shipped off to exclusive restaurants at Bellagio Las Vegas or in New York City.

As the crow flies, de Lancellotti Vineyard is precisely a mile east of Trisaetum (one must simply pass over the patron saint of Oregon biodynamic vineyards, Brick House, to get there). The twenty-seven-acre de Lancellotti farm is planted with seventeen acres of vines, which, at present, produce a mere five hundred cases of wine annually, with the remainder of the harvest going to the extended family's winery, Bergström.

Which is next door.

The Bergström family domain was originally financed by Dr. John Bergström, Paul's father-in-law; it's led by Kendall's smart thirty-something brother Josh, who traveled to Burgundy just out of college and completed the two-year Viticulture, Enology and Wine Business degree at Lycée Viti-Vinicole de Beaune in just one year, in French. While in France—where, in addition to studying, he met and married his wife Caroline, in a whirlwind romance—Josh Bergström learned about biodynamic viticulture. His school curriculum covered it in passing, but more important, he noticed that the prestigious *domaines* of Leroy, Leflaive, and Romanée-Conti were doing it, and people were talking about it. "I was really intrigued," he recalls. "Why was it that these biodynamic producers just happened to be the best names in the world?"

After he returned home, Josh and his family planted the fifteen-acre Bergström Vineyard in the Dundee Hills in 1999. After a traumatic early trial with an herbicide—which nearly killed the entire fledgling vineyard—the young vigneron swore off chemicals and began to farm biodynamically in the spring of 2000. Two years later, he harvested his first fruit and began making wine. Soon after, the scores started coming in: a 94 from the *Wine Spectator* for the first vintage; a 95 the next.

The accolades have allowed Bergström to raise prices— somewhat—for the premium-quality wines that he produces. "Price is funny when it comes to Oregon," says Bergström. "We are making efforts here that are world-class. We have some of the lowest yields in the state, and we implement a very expensive farming system. We pay our team very well. We have great packaging, and we buy nothing but the best French oak barrels. If we followed the lead of Napa, Burgundy, or Bordeaux, our wine would be $200. Our most expensive wine is $85, and that's still a tough pill for people to swallow."

Bergström, too, is an Armenier client. Its three estate vineyards— including de Lancellotti—are Demeter certified. "When we first started doing biodynamics, I don't think it really helped at all in terms of marketing. But nowadays, it's really changing. The movement across the nation is really taking off," Josh Bergström says. "Now there is so much wine being made across the planet. Biodynamic certification tells the customer it's a unique wine from a unique farm, where grapes are grown in a holistic way. It's not just mass-produced stuff."

So don't bother looking for it in the wine aisle of your local supermarket. Because this isn't any old organic wine.

But if you're lucky, you might—just might—find a bottle of Bergström at Whole Foods Market.

A 2007 Rolling Stone article about the reunion tour of The Police shows two sides of bandleader Sting. On the one hand, he's an emperor, "fit and blond," with a "kingly hotel suite," who, according to guitarist Andy Summers, runs the band like "an ego-cracy." On the other, he's driven and detail-oriented, demanding that Summers rehearse the same three notes with him for half an hour.

Is Sting a pasha or a perfectionist? Both, apparently. But do we have a problem with that? "The Police were together for only seven years, but the band's fusion of punk and reggae influenced an entire generation of performers who followed," the sidebar to the article states.

To be elite is to open oneself up to criticism. To pursue perfection is to be imperfect. Sting came to be one of the best musicians on the planet because he is exacting. Because he is exacting, he is considered to be an asshole.

One can see why Sting might want to farm his Tuscan vineyards biodynamically. It's the most difficult, most exacting way to go about making wine. It is a pursuit that only a perfectionist would have the patience to put up with.

It's easy, too, to see why high-end American biodynamic winegrowers are often criticized for "bragging about being BD," as one commenter notes on the blog *Biodynamics is a Hoax*. The implication is that biodynamics is merely a bragging right, a way to claim an association with European producers and a reason to tack a few more dollars on the bottle price. It's tempting to lump American biodynamic wines together with the expensive biodynamic skin creams at Whole Foods. Are those prices truly justifiable?

But when I ask the most glamorous winegrowers of Oregon why they're pursuing biodynamic viticulture, they tell me that just growing the best grapes and making the best pinot noir they possibly could wasn't enough. They're on an Arthurian quest to achieve the impossible: to create a distillation of both heaven and earth.

CHAPTER TEN

The Green Factor

> The all-essential causes of what happens on the earth
> do not lie outside the human being; they lie within
> humankind.
>
> —Rudolf Steiner

In 1797, William Tuttle purchased three hundred acres of land in Freeport, Maine, and began to raise cattle and crops on it. For the next century, the Tuttle family continued to farm this piece of property until something strange occurred: patches of sand appeared on the earth.

Today, the farm is gone. In its place, massive sand dunes rise as high as fifty feet, engulfing buildings and trees. The property has become a tourist attraction, called "The Desert of Maine." As tour guides explain to some thirty thousand visitors each year, poor farming practices—the removal of trees, overgrazing by sheep, and the failure to rotate crops—eroded the topsoil of the Tuttle farm to the point where the glacial silt underneath was exposed. Freed from its fertile tether, the sand erupted and flowed like slow-moving lava until it covered the entire estate.

What happened to the Tuttle farm is called desertification, and it continues to happen every day, all over the world. The removal of trees and bushes destabilizes the soil structure. Chemical additives such as fungicides break down the humus that feeds plant life. Overgrazing, overtilling, and strong winds strip this lifeless topsoil from the earth's shell, leaving nothing but sand behind. On a grand scale, desertification results in environmental catastrophes such as the Dust Bowl, familiar to us from those iconic black-and-white photos of farms engulfed by sandstorms. This tragic transformation of farmland to desert is credited with causing the migration of some 2.5 million Americans from prairie states to coastal areas between 1930 and 1940.

Oregon offered a forewarning of this disaster some three decades earlier. On a June Sunday in 1903, a two-story wall of water raged through the prosperous farming town of Heppner, destroying everything in its path. The Heppner Flood was the deadliest natural disaster in Pacific Northwest history. Scientists later concluded that the tragedy had been caused partly by human folly. "Soil erosion was a major reason why the flood killed 245 people," says Joann Green Byrd, author of *Calamity: The Heppner Flood of 1903.* "And most of that erosion was caused by overgrazing and plowing of the land."

According to Byrd, intensive farming had left the hills above Heppner rock-hard and devoid of vegetation. Lacking absorbent soil, the hills acted as a conduit, channeling the water from a fierce and sudden rainstorm down into the canyons above the town. "The water became compressed in these canyons and began picking up boulders the size of refrigerators, fence posts, cows, horses—everything that got in its way. It carried these things and used them as bludgeons. This tremendous amount of water would not have been able to run off those hills if the vegetation had still been there," says Byrd.

More than a century later, farmers know to work the land more carefully. Still, "there is concern about that occurring again," warns Jay Noller, a professor of landscape pedology at Oregon State University. "That's the whole reason we have soil conservation districts. We have government and non-government organizations that work on this year-round."

Of course, in arid eastern Oregon, it's easy to identify soil that might enable a flash flood: it's dry, brittle, and pale. In the Willamette Valley, we have tons of water—it once famously rained for thirty-four days straight—and we have steep slopes. But the climate isn't so conducive to flash floods. The rain comes down in a light drizzle and the earth tends to be spongy and verdant with groundcover.

When a Willamette Valley winegrower worries about the consequences of his farming practices, desertification is not something that immediately springs to mind. Because the color of poor soil health here is far more insidious than the obvious ivory of sand: often, it's an unremarkable green.

"Moss is a primary colonizer, like lichen," says Dan Rinke, viticulturist and winemaker at Johan Vineyards in Rickreall, Oregon. "It is the first thing you see growing on a rock with no microbial life. Seeing moss under vines? That makes me sad." A bearded bear of a guy

approaching his mid-thirties, Rinke wears a hooded sweatshirt and baseball cap and speaks with the folksy inflections of a Midwesterner. In a previous life, he worked in a wine shop, then as a sales rep for a Milwaukee wine distributor, where he became familiar with the wines and philosophies of Nicolas Joly and Michel Chapoutier. It was on the advice of Chapoutier himself that Rinke enrolled in 2002 at California State University, Fresno, for a degree in viticulture (rather than enology, as he had originally planned).

"At that point in time, organic and sustainable farming were just starting to become more popular. It was something that we felt as students that we should have as part of the curriculum," Rinke recalls. But instead, all one thousand acres of demonstration farmland at Fresno State were devoted to conventional agriculture. Frustrated, Rinke and a group of friends founded a club called Students for Environmentally Responsible Agriculture (SERA), which planted an organic plot on the campus farm in 2004. Rinke then formed a biodynamic study group at school and apprenticed himself to Chris Velez—vice president of the Biodynamic Farming & Gardening Association and co-owner of Stella Luna Farms in Auberry, California—to learn how to make and apply the preparations.

After working as a winemaker and vineyard manager for two California biodynamic wineries, Rinke moved north in 2007 to oversee winemaking and viticulture at Johan, a new winery near Salem. The owner wanted to farm organically, but Rinke convinced him to pursue biodynamic certification.

After two years of farming biodynamically at the edges of urban areas, it was, for Rinke, a no-brainer to do so in the Oregon countryside. "Biodynamic farming is so much easier up here: I've got close friends who raise organic grass-fed dairy cows and organic free-range chickens. I can get eggshells and really good, high-quality manure for barrel compost."

Rinke's interest in biodynamic viticulture was driven by a pursuit of purity. As he sees it, wine with *terroir* comes from grapes that express a sense of place. To add chemicals to that place clouds that expression.

But it's also a moral decision. Because while he's a purist when it comes to wine, he's also an activist when it comes to the environment. "I'm farming a piece of land that will be handed down for generations and I want to keep it as pure as possible," Rinke says. "It makes me happy not to see moss under the vines."

Of course, there are more mainstream ways to go moss-free than biodynamic agriculture. Approximately 30 percent of Oregon vineyards are certified by one of the state wine industry's many sustainable-winegrowing initiatives—LIVE (which includes Salmon-Safe certification), organic, biodynamic, or the all-encompassing Oregon Certified Sustainable—with countless others practicing eco-friendly farming without certification papers.

LIVE (for Low Input Viticulture & Enology, Inc.) certification requires that at least 5 percent of a vineyard property's total acreage be set aside for wildlife but allows for use of a wide variety of fungicides, herbicides, and pesticides, if applied carefully and in moderation, in cases of documented need.

In addition to requiring buffer zones, USDA Organic certification also mandates crop rotation or use of cover crops, composting, and mechanical rather than chemical pest and weed control. That said, a lengthy list of synthetic soil amendments—toxic substances such as boron, which you wouldn't want your dog to accidentally eat—are permitted.

Demeter doesn't allow for these additives, making biodynamic the most stringent certification out there. But practitioners bristle at the oft-tossed-out "über-organic" label. "We're all trying to farm responsibly," goes a common refrain, "so please don't place us on a continuum and pit us against one another."

The vignerons who participate in all of these programs dutifully seek independent third-party certification that their vinetending practices limit human exposure to pesticide sprays. And they are forever mindful of the basic fact that vineyards tend to be located on hillsides.

Compared with the wind-buffeted plains of arid eastern Oregon, it might look like a Garden of Eden here, but as OSU's Jay Noller puts it, "Wind is one fluid. Water is another. That's the major concern in the Willamette Valley: soil erosion." Erosion leads to runoff, which ends up in rivers, where it can harm native salmon species and other wildlife. From there, the reverberations can be felt throughout the ecosystem.

And so the vinetenders of Oregon are conscientious farmers. But how much impact can their conscientiousness really have?

There are sixteen million farm acres in Oregon; of these, just 0.12 percent, or 19,300 acres, are vineyard. Of the wine-centric Willamette Valley's 1.73 million acres of agricultural land in 2009, vineyards only accounted for a measly 14,556 acres, or 0.84 percent of the total. By contrast, the grass-seed farms on the valley floor took up 495,000 acres, or nearly 30 percent of the total. So, what difference could it possibly make if 30 percent of Oregon winegrowers have eco-certification? The wine-grape sector is a tiny slice of Oregon's agricultural pie.

Geographically, yes. Economically, no. According to the Oregon Seed Council, the overall economic impact of the grass-seed industry was more than $2 billion in 2008. According to a conservative estimate commissioned by the Oregon Wine Board, the economic impact of the Oregon wine industry was $1.4 billion in 2004. Comparing these numbers—which are, remember, four years apart—the importance of the wine industry becomes clear. Because, acre for acre, the wine industry is, at the very minimum, twenty times more profitable than the grass-seed industry.

And if an industry can be this profitable while preserving our natural resources, it should cause consumers and lawmakers to sit up and take notice. Granted, grass-seed farms are invaluable contributors to Oregon's economy. But they also take a toll on the environment: they must be annually cleared (using either controlled fires or a combination of machinery and chemicals) and replanted, whereas hillside grapevines live for decades in the same spot.

Oregon's winegrowers have been integral in the state's historical record of land preservation, as well. "When the wine industry was just beginning to take root in Oregon in the early 1970s, most of the land now covered with vineyard was considered to be secondary low-value crop land," recalls Pat Dudley, co-owner and co-founder of Bethel Heights Vineyard in Salem and a member of the Oregon State Board of Agriculture. "Most was rocky hillside with uneven terrain and shallow soil that would have been abandoned and become residential or commercial development if the wine industry had not taken over. Our vineyard was slated to become a mobile-home park when we purchased it."

But the argument for the green impact of *biodynamic* viticulture in Oregon is less convincing: as this book went to press, a mere nine hundred and nine Oregon vineyard acres were certified biodynamic, accounting for only about 5 percent of Oregon vineyard land. (By my most recent calculations, non-certified BD vineyards accounted for an additional 365 acres; even if you include this, you're still only representing about 6.5 percent of total vineyard land in Oregon.)

It's easy enough to convince privileged collectors who invest in fine French wines of the charms of biodynamic viticulture. But it's more difficult to see how biodynamic vineyards—which account for a mere speck in the sea of American farmland—could be making a difference in the battle we're currently losing against global warming, pollution, desertification, and other environmental threats. "The biodynamic movement is so small as to be insignificant," admits Sam Tannahill, director of viticulture and winemaking for the wineries A to Z, Rex Hill, and Francis-Tannahill. "But what we're trying to do is not only sustain, but to heal: to increase the health of the earth."

It's a tall order.

But for Glenn McGourty, it's not an impossibility.

For McGourty, a winegrowing and plant-science advisor with the University of California, Davis, a solution to some of our planet's biggest problems is simple, elegant, and right in front of our eyes. It's a matter of growing cover crops instead of expending untold amounts of kilocalories and dollars on the production and purchase of fertilizer. It's a matter of piling potassium-and-nitrogen-rich pomace and yeast lees onto a compost heap rather than paying thousands of dollars a year to have this refuse hauled away to a landfill.

We might call them biodynamic or organic practices, but really, these are acts of simple thriftiness. McGourty calls them "really good agronomics." And he asserts that, implemented on a larger scale, these practices could be a part of the solution to the global warming problem. "It's estimated that 30 percent of the carbon dioxide in the atmosphere has been lost from the soil due to poor farming practices," says McGourty. "If we look at what the Midwest looked like a hundred and fifty years ago, there would have been 20 percent more topsoil. The carbon from that soil ended up in the atmosphere and our water."

As McGourty sees it, farmers might be able to reverse this trajectory through careful composting and crop rotation, which not only preserves a farm's own carbon, but even sequesters carbon from the atmosphere. A self-contained, recycling-reliant, "carbon-cycling" farming system ensures that any carbon produced on site stays on site instead of being released into the water and the air; the addition of cover crops provides even more carbon-dioxide-drinking green material.

Over a span of ten years, the carbon content in the soil of such a farm can—under the right circumstances—increase tenfold, says McGourty. "That is carbon that otherwise would be floating around in the atmosphere," he asserts. Here's how it works: During photosynthesis, plants such as cover crops, trees and, for our purposes, grapevines, consume carbon dioxide from the atmosphere and convert it into carbon. Later, cover crops are mowed, leaves and fruit drop to the ground, and compost is sprinkled on top. The microorganisms in the compost, joined by earthworms, munch on all the carbon contained in the green matter and break it down into carbon-rich humus. Thus, the carbon that was once in the atmosphere is now stabilized in the soil, where it nourishes new plant life instead of heating the planet.

It's easy to tell if McGourty's environmental solution is working. One need simply look at the soil: The more carbon in the humus, the richer and deeper the color of the soil. Sometimes it's even black—the opposite of ivory-hued sand.

Now for the deep shit.

For today's fooderati, animal-confinement operations are the poster children for bad farming, where animal cruelty is just a part of doing business, and the end result of this business is our own poor health: the engendering and spreading of viruses and delivery of unnecessary antibiotics into our bodies, hampering our ability to fight diseases. The environmental impacts, too, can be catastrophic. They include habitat destruction and land degradation. And, perhaps most significantly, many factory farms produce as much sewage as cities, but leave this sludge untreated, imperiling our water and air. To read writers such as Michael Pollan and Eric Schlosser is to be convinced that the earth is on the verge of drowning in a cesspool of cow manure.

"Have you ever visited a dairy? They're really hard-core," the winemaker Sam Tannahill confides. "When we first started farming biodynamically, we went around looking for places to get organic manure to make our compost. Even among those that said they were organic, there was only one farm where the cows actually got to go out to graze *and* they didn't separate the manure into liquid and solid. They just shoved it all to one side, which was perfect because we could compost that."

For Tannahill, the issue of environmentalism always comes back to the cow. "Consumer expectation is for cheap produce and cheap meat, so you have to have these efficiencies in order to achieve the price point that people are looking for," he says with a shrug. "Through subsidies and chemical agriculture, we have led people to expect and demand cheap food, and in some ways that is great because we are able to feed so many people so cheaply. But what is the real cost? What has the toll been on our topsoil, and on global warming?"

Tannahill, Rinke, and other biodynamic practitioners think that their style of farming offers a solution to this problem: composting. Because biodynamics, more than any other style of agriculture, is dependent on the availability of cow manure. And by composting this manure into a rich humus amendment for nutrient-deprived soil, it turns shit into gold. For them, biodynamic farming isn't just a reactive defensive mechanism for protecting a crop from harm; it's a proactive strategy that aims to nourish the crop and protect it as well as eliminate waste, decrease pollution, and enrich soil.

Scientists are often hesitant to talk about biodynamic agriculture. After all, it's an untested, unproven, spiritual way of farming. But for Dan Sullivan, a soil scientist with Oregon State University, biodynamic composting is one of many ways that the waste created by confinement agriculture can be put to good use. Of course, Sullivan points out, there are plenty of responsible Oregon farmers who already integrate their cattle waste back into the land: "It is perfectly acceptable to apply manure directly to cropland and grow crops. This is something that has been done since the start of farming."

However, direct application poses a health hazard. When fresh manure is sprayed onto crops, particles can linger in the air and be inhaled. Composting takes time and manpower, but, says Sullivan,

"The heat in the process kills any pathogens that might be transmitted to people or the environment." It also reduces the volume of the manure, making it easier to apply to whatever part of the farm needs it most. "There are a lot of different ways you can use this material," Sullivan says. "You can use it to supply nutrients, to improve how water is stored in the soil, to improve the soil tilth. I would look at biodynamics as kind of a value-added use for manure."

Does Sullivan see all the cow manure in the country being converted into biodynamic compost? Not in a million years. But he does see BD composting as one of many ways that a new green economy could create value out of a raw material that is currently creating havoc. Manure also can be converted into biogas; it can be burned to produce electricity. "There are all kinds of things you can do with this," he points out. "It's just a matter of what the enterprise data sheet looks like for your product. Does the return on investment work out for you?"

For a high-end winery, biodynamic composting is a worthwhile investment. Because for winemakers like Sam Tannahill, biodynamic farming makes for a compelling story. And it is this story that sells wine. But the point of the story, says Sam Tannahill, is not to raise prices. It's to farm responsibly, and to inspire others to do so. "Steiner said that by the turn of the century we would have our tables laden with fruit and vegetables and meat that all would look gorgeous but have no nutritional value whatsoever. And that's really turned out to be true. And I think it's what people like Michael Pollan are getting at," Tannahill reflects. "When you see reports of pesticide residues being very high in wines, what is the true cost?"

With thoughts like this in mind, Tannahill and his co-winemaker wife, Cheryl Francis, began farming their own ten-acre Pearl vineyard biodynamically in 1999, when she was winemaker at Chehalem and he was at Archery Summit. Under a one-hundred-and-sixty-year-old maple known locally as the Baptist Tree are an herb garden for biodynamic prep making as well as groves of cherry, plum, walnut, apple, fig, olive, and hazelnut trees, and blackberry and raspberry bushes. The skulls of Ocho and Nueve, the couple's first two steers, hang like talismans over barn doors.

Francis and Tannahill don't live at their vineyard anymore—they have since moved to a half-acre homestead in Portland, where their three small children can attend school. Still, their new urban yard is overgrown with raspberries, blackberries, quince, red and white currants, herbs, and vegetables. It's as though everything this couple touches turns to green.

And that's on paper as well. As co-owners of A to Z, one of the two largest wineries in Oregon, they produce more than a hundred and twenty-five thousand cases of wine annually. Rex Hill, Francis-Tannahill, and another affiliated label, William Hatcher Wines, produce some ten thousand cases altogether. Combined, the operations currently farm ninety-three acres biodynamically (of which 35 acres are Demeter certified), with plans to increase that number.

When Sam Tannahill compares the act of pushing young vines to produce quality fruit in their second year to that of pushing a yearling calf to breed early and make milk, it's clear that he knows how most large winemaking businesses are run. And that he's chosen not to run his own big operation this way. Despite the challenges of running such a large winery operation, for example, he insists on making the biodynamic preparations on-site rather than purchasing them pre-made.

How big an impact are people like Sam Tannahill having on the current environmental crisis? If you consider that they farm 5 percent of the total acreage of Oregon vineyards, very little. But if you consider that the passion of people like Sam Tannahill convinced at least one journalist to write a whole book about biodynamic winegrowing in Oregon, maybe they *are* making a difference. "People are trying to make a change, and I think biodynamic winemaking has an opportunity to do that. Because wine—unique among all agricultural products—has a built-in platform to tell a story, to get the message out," Tannahill says.

Leigh Bartholomew and Patrick Reuter are master storytellers. The UC Davis-educated couple have worked in wineries all over the world, including a memorable stint in France where, in Alsace and Burgundy, they were beguiled by biodynamics, sneaking onto top estates—such as Domaine de la Romanée-Conti and Domaine Leroy—to peek at the vines and observe biodynamic viticulture in action. Returning to Oregon, they hatched a plan to grow tempranillo and syrah.

Reuter, who wrote his master's thesis on American *terroir*, felt that the poor soils of the Columbia River Gorge were ideal for nurturing tempranillo vines. So he and Bartholomew devoted their weekends to searching for vineyard sites along the liminal stretch of steppes where the rainforest surrounding Mt. Hood gives way to the arid landscape of eastern Oregon. In the Willamette Valley, the average annual rainfall is thirty-five inches. But in Mosier, on the Columbia River Gorge, it's seventeen inches.

Farming is tough in this region, where arable land nestles above cliff faces and under national forest, and where *mistral*-like winds charge down the Gorge at a consistent strength of thirty-five miles per hour. And where there's barely any topsoil, thanks to the great Missoula Floods, which periodically tore through the Gorge some thirteen to fifteen thousand years ago, ripping up layers of earth and depositing them on the floor of the Willamette Valley.

"We spent a bunch of time hopping fences and digging soil pits," Reuter recalls. "I had all sorts of stories in my mind—what I would say if I got caught. We finally identified fifteen properties and sent letters to all the owners to ask if they were interested in selling. We were working with a real-estate agent who thought we were crazy. But we wanted a certain soil profile, slope, and elevation." The ploy worked: one property owner was willing to sell. A former dry-farmed cherry orchard, the land was nothing more than a field filled with stumps when the couple purchased it in 2001. When they walked, puffs of dust clouded around their feet; when they picked up fistfuls of soil, it blew away like sand.

Undeterred, the couple convinced Bartholomew's parents, Liz and Glenn, to retire on the property, which they named Three Sleeps Vineyard; they named their wine label Dominio IV after the four partners. And the family rolled up their sleeves and went to work: damming a couple of winter springs to lessen erosion and collecting the water in ponds to create wetland habitat. Clearing brush to deter forest fires.

Slowly, out of the dry land, they have begun to build a garden. There are rows of fledgling vines. A volunteer apple tree has been espaliered to the deer fence. The soil that once slipped through their fingers now stays packed in their palms in solid, moist clumps. Reuter and Bartholomew dream of planting fruit trees, husbanding animals, building guest houses, hosting groups of inner-city children. But for now, while Bartholomew's day job—vineyard manager for the prestigious Willamette Valley winery Archery Summit, Sam Tannahill's old stomping grounds—keeps the couple living in McMinnville, they are content to drive out to their vineyard on weekends and let their two young boys run loose while they tend the property.

Because their soil was so degraded when they bought the estate, and because they were so impressed by the biodynamic wines they tasted in France, and because they are concerned about the environment, the couple has Demeter certified Three Sleeps Vineyard. And they tell their story whenever they pour their sumptuous tempranillos, syrahs, and pinot noirs.

"Roundup is very tempting and the hoe is a lot more work," says Reuter. "But if you tell people that instead of spraying a fungicide, we use stinging nettles to the same effect, consumers like to hear that: that there are no chemicals involved in the process. People can see the practicality of that, and they like that we're farming that way."

Someday soon, Reuter plans to plant a grapevine labyrinth modeled after the graphic motif inlaid into the floor of the Chartres cathedral. Divided into four quadrants, one for each season of the year, and with lunettes symbolizing the phases of the moon, its rows will point straight at the rising solstice sun. It will be a living calendar, a maze, and simply an interesting way to configure vine rows. It will make for a good story. One can imagine, decades from now, the Dominio IV labyrinth attracting a certain sort of eco-tourist, someone who might have visited the Desert of Maine ... had it not been boarded up.

Because, you see, word has it that the sands of this anomalous desert are—ever so slowly—being subsumed by the surrounding forest.

CHAPTER ELEVEN
The Naysayers

The language of Biodynamics, with its references to the
alignment of the planets, undefined life forces, and the use
of bizarrely fashioned preparations, seems totally at odds
with a rational, scientific world view. As a consequence,
most scientifically literate people have dismissed
Biodynamics altogether; alternatively, they have regarded
Biodynamic practices as affectations, and explain any
benefits merely in terms of increased attention to vineyard
management.
—Jamie Goode, *The Science of Wine: From Vine to Glass*

"So we're walking through the vineyard and we point out that we have
these gopher problems. And this French consultant says, 'Theess is
what you need to do with zee goph-airrr: You cut off its head and put
it on a spike at zee end of your vine-yahrd. And zat negative enay-
hrgee will scare zee goph-airrs away.' And I'm like, 'What is this,
Spartacus? I don't want *Spartacus* in my vineyard! This is awful!'"

Alex Sokol Blosser is tall, with a friendly face and generous
dimples. He's energetic and speaks quickly. He is disarmingly frank
and assiduously funny: he easily could do voice-double work for
the comedian Seth Rogen, and would have no trouble coming up
with material (his version of the shtick would be cleaner, though).
So when Alex Sokol Blosser compares biodynamic viticulture to a
gory Kubrick film, one can't help but laugh. This is a guy who calls
his elegant dry rosé "our seasonal beer"; his promotional vineyard
videos on YouTube feature the recurring character of a silent ninja
cellar rat, for no apparent reason other than the fact that it's outré.

But when it comes to running his family's winery, Alex Sokol
Blosser is dead earnest. He has served on just about every position
on just about every board in wine country, including president of
the Willamette Valley Winery Association. He was the impetus
behind the Willamette Valley's division into quality-oriented sub-
appellations, starting with Sokol Blosser's own Dundee Hills
American Viticultural Area.

And the Sokol Blosser commitment to ecologically sensitive winegrowing goes without question. In 1996, the Salmon-Safe eco certification launched with a press conference at the Sokol Blosser tasting room; in 1999 Sokol Blosser was among the first group of Oregon vineyards to be LIVE certified. Every year, the winery releases a brutally honest sustainability report, celebrating small victories and reproaching itself for missteps.

Certified-organic vineyard? Check. First winery in the world to achieve U.S. Green Building Council LEED certification? Check. Renewable wind power? Check. Solar panels? Check. Biodiesel? Check. Active in the Oregon Natural Step Network? Check. Check, check, check.

Susan Sokol Blosser, Alex's mother, was the original driving force behind all of these achievements. In 1971, she founded the winery with then-husband Bill Blosser; since then, she has been among the most zealous of the Willamette Valley's green-minded vintners. For example, she spearheaded cover-crop trials in collaboration with Oregon State University and the Yamhill Soil & Water Conservation District as far back as the early 1980s. But by the late 1990s, the Sokol Blosser family began to see the health of their older vines declining. Could their vines, like longtime smokers, be suffering from the effects of decades of chemical dependence? The family began farming organically in 2001 to find out if a more natural approach could invigorate their vineyards.

"That wasn't that long ago, but there still weren't that many people who could tell us what the hell we were doing," Alex recalls. "I would call wineries down in Mendocino County, where there is the highest concentration of organic vineyards in the U.S. I would talk with them and realize that I knew more than them, that this was a waste of my time." He mimics holding a phone to his ear, then staring at it in exasperation and hanging up: "Hey, thanks, buddy."

Meanwhile, Susan had joined the Oregon biodynamic study group led by the eccentric Andrew Lorand. "That was something that really piqued my mom's interest, because there were a lot of incredible wines coming out of Burgundy and Alsace that were biodynamic," Alex recalls. "Biodynamics is really, I've gotta say, a lot sexier than organics."

Until you start practicing it yourself, and realize it's the opposite of sexy.

"Apparently the shape of some pregnant woman's belly made for the perfect vortex, the swirling and the spinning ..." As he traces a spiral in the air with his finger, Alex Sokol Blosser sputters with laughter at this recollection. "So, basically, we would mix all our compost teas and prep sprays in this belly-mold thing. We used this stuff in this block"—he gestures at the row of grapes behind him— "and we had the worst outbreak of mildew ever. It was insane." That was 2001, not a particularly bad mildew year for the Willamette Valley. But in Sokol Blosser's Bluebird Block, the mildew got so bad that Alex and Susan had no choice other than to use Rally (the sterol-inhibiting chemical fungicide) to save that section of their vineyard from total ruin.

After the Lorand group dissolved, Susan Sokol Blosser and a few other winery owners hired the consultant Philippe Armenier to periodically travel to Oregon and meet with each of them individually. Alex went along with Armenier's advice ... for a time.

Then there was the Great Vole Infestation of '05, when the consultant advised them to mulch their vines with straw bales. "To the voles, these looked like condominiums," Alex recalls ruefully. He starts to count the number of his neighbor's vines he and his mother killed during this episode, trailing off at eight. "And we killed hundreds of our own vines. Hundreds. Oh, my god. My mom was terribly embarrassed about that."

Now that his mother is semi-retired from the family business, Alex, who serves as co-president of the Sokol Blosser winery with his sister, Alison, has embraced Susan's drive to make the family winery as sustainable as possible. "We don't want to be perceived as one of those businesses for whom it's all talk," he says. "We want there to be some stuffing in the pillow. With the sustainability report, we're saying, 'Hey, we don't do everything right, but we do have a triple bottom line: people, product, planet.'"

And so Alex is warehousing bundles of used plastic until the market for recycling the stuff turns around, because he can't bear to throw it away. He sets aside vineyard space for an employee vegetable garden. He uses pressure-sensitive labels printed on plain recyclable backing rather than the traditional backing, which isn't recyclable. But the biodynamic viticulture has fallen by the wayside. "The direction I want to go is I want to be the best organic farmer out there," he says. "I have zero tolerance for mildew. I have zero tolerance for rot. Organics are very much like conventional farming, except that you cannot use synthetic chemicals. And that little law means that you have to focus on the soil. You've got to really be in tune with the health of your soil.

"I love talking to people who farm biodynamically. They are very passionate about it, very into it. You talk to Rudy Marchesi, you talk to Josh Bergström, and pshew! They love it! Good for them. It works for them. But it doesn't work for me or for our team," Alex confides. "If you're doing it halfway, you're really not doing it. I don't believe it in my heart; therefore I couldn't make it happen."

It's easy to see why a winegrowing family like the Sokol Blossers—earth-loving, quality-oriented pinot pioneers—would have been attracted to biodynamic viticulture. It's regularly referred to as "the most natural" way to farm grapes; it's a farming style that is inextricably intertwined with the "natural wine" movement.

To which the jaded vigneron replies, "What do you mean by 'natural?'"

"Originally, wild grapevines were either male or female, and only half of them would produce fruit," points out Steve Doerner, winemaker at Cristom Vineyards in Salem. After thousands of years of hybridization, "today, the vines that we have bred to grow commercially are all hermaphrodites, so they're all fertile. All of viticulture is man-made."

Grapevines aren't grafted onto rootstock in the wild; instead, native cultivar species grow from seed, and climb the trunks of trees. Their clusters are loose, their berries small. To breed plants for tight, thick fruit clusters, graft them onto rootstock, and then line them up side-by-side in neat vineyard rows, supported by trellis wire, is to thumb one's nose at nature and taunt fate. Monoculture is an open invitation to pestilence. This is not natural.

The argument in favor of "natural" farming techniques is reminiscent of the conflicting child-rearing philosophies *du jour*. In one corner are the "attachment parents," who carry their babies around in slings, nurse them at any moment, and sleep with them at night. In the other are parents who believe that cribs, strollers, and bottles are best for their kids. There's no conclusive evidence that either style of parenting is better or results in more successful, well-adjusted children. But the attachment camp's argument that this method is superior because it is practiced by primitive societies seems suspect: so should we all also walk around with bared breasts, practice polygamy, perform female genital mutilation on young girls, promote infanticide, and sell our children into slavery?

Likewise, why aren't the "natural" vinetenders riding donkeys, or walking through the vineyards barefoot and doing all their work by hand? Surely mechanized equipment such as tractors and ATVs aren't "natural."

"People like to have a unique farming system that is not necessarily conventional; they want to do something different," observes Patty Skinkis, assistant professor of horticulture at Oregon State University. "Biodynamic agriculture reverts to a time in which we didn't understand as much." A viticulture extension specialist, researcher, instructor, and director for the American Society for Enology and Viticulture, among other things, Skinkis has seen every type of vineyard-management system at work. Although she has seen successful examples of biodynamic viticulture, "A lot of what has unfortunately occurred is that a lot of people who have not as much education in vineyard management and vine physiology want to go this route," Skinkis says of BD. "People who have been trained professionally have different thought processes."

Which is another way of saying that if you ask most professional vineyard managers what they think of BD, they'll roll their eyes and groan. I found it difficult to find a single contract vinetender who was willing to be quoted on the record on the subject of BD; as one told me, "A lot of vineyard guys get it shoved down their throats by the owners. They think it's stupid, but they've got to do it."

"Farmers are risk averse. For good reason," explains vineyard manager Stirling Fox, whose Stirling Wine Grapes company manages three hundred acres of vineyard for eighteen different Willamette Valley clients. "If you farm conventionally, you have enough weapons in your arsenal to routinely keep your vineyard disease-free." But, unlike many of his colleagues, Fox welcomes the challenge of farming biodynamically for three of his clients, Evening Land Vineyards, J. Christopher, and Scott Paul.* "I would say it's about 10

---

* None of these wineries' properties were Demeter certified during the time that I was researching this book.

to 15 percent more expensive than LIVE (sustainable viticulture),"
Fox admits. "But you're out in the vineyard more. You're giving the
vineyard more attention, and the vineyard responds to that. You
have to manage a very clean vineyard. You can't afford to blow it
because the risk is bigger. So you put in more time and more effort.
We think that gets you a better result."

Sometimes Alex Sokol Blosser asks himself if he was going about
biodynamics in the wrong way. Was he seeking the advice of the
wrong consultants? Should he have studied Steiner more thoroughly? But then he thinks about all that ... poop: "I don't want the
vineyard guys to do anything I wouldn't do. And I don't shovel shit.
The guys would go home and their wives would say they stank. So
we stopped. I'm not doing that to my guys." Cow manure doesn't
just stink; it's also heavy and awkward. In most cases, it must be
loaded on the back of a truck and lugged to the vineyard before it
can be dispersed, using tractors.* All of which consumes fossil fuels.
How green, really, is that?

As Willamette Valley vineyard manager Joel Myers points out, the
most efficient way to compost is simply to allow pruned cuttings
and mowed cover crops to decompose out in the field where they
lie: It's "growing compost in the vineyard rather than importing it,"
explains Myers. "If you do nothing but cut your grass and chop your
brush, you're going to be OK."

A study published in the *American Journal of Enology and Viticulture*
in 2008 compared the effects of six different types of organic soil
amendments in a Loire Valley vineyard over a twenty-eight-year
period. "Results showed a trend toward a favorable influence of
crushed pruned wood treatment on vine behavior," writes the
author, praising a more passive composting method such as the one
Myers describes.

A plainspoken teddy bear of a guy with silver hair and round
spectacles, Myers has been a key player in the Willamette Valley wine
industry since 1968, when—just ten years old—he helped David and

---

* As the biodynamic evangelist Peter Proctor has shown, it's easy to
practice biodynamic agriculture in India, where cows are ubiquitous
(for more on this, see the film *One Man, One Cow, One Planet,* www.
onemanonecow.com). But does it make sense in a place where cows are no
longer a part of every farm?

Diana Lett to plant their historic Eyrie Vineyards. Today, in addition to his vineyard-management work, Myers bottles his own wine under the Vinetenders label. He has grown organic vegetables for a living; he has studied grapevines at the Agroscope Changins-Wädenswil research station in Switzerland; and he has managed the vineyards of the Valley's top estates, counting among his clients biodynamic practitioners such as Doug Tunnell and Laurent Montalieu. He is one of the most respected viticulturists in the Valley.

For Myers, labor equals time equals money. And in his experience, farming organically or biodynamically is more expensive *and* carbon-intensive than the sustainable style he prefers. Tractor cultivation, he points out, consumes a lot more fuel than does a guy on a four-wheeler spraying a single quart of glyphosate herbicide such as Roundup; it's also about $500 per acre more expensive: "The growers are saying, 'Well, I'm going to spray a little Roundup right now because I can't afford to hoe.'

"And I don't consider hand-hoeing to be sustainable," Myers continues. "It's a terrible job. I did a lot of that as a kid out in the fields. You are relying on peoples' backs and blistered hands. You see wealthy vineyard owners sitting in a nice restaurant saying how green they are because they have forty people out there weeding. Well, I don't agree with that because they aren't out there. It was a hundred degrees out there last week, and I don't think that's fair.

"We've had several fads over the last thirty years: trellis fads, spacing fads, clone fads, irrigation fads, rootstock fads. Now we're having this green fad," Myers concludes. "Now it's, 'How green am I and how green are you?' These fads come and they go. We're just in the middle of this one. We're going to phase out of this one because this one costs a lot."

Or, even if the "green fad" doesn't phase out, it's not clear that biodynamics will maintain its claim as the greenest of the green. Holistic farming methods are being embraced by the mainstream so quickly that many aspects of BD farming are looking less and less special. According to recent reports on National Public Radio and in *The Wall Street Journal*, the year 2010 saw an increase in the number of parks and private properties making use of "free-range landscaping" from goat-powered services like RentaRuminant.com. And states like Ohio and California have recently moved to ban the tight confinement of animals in farming operations.

Threemile Canyon Farms in Boardman, Oregon, is a ninety-three-thousand-acre ranch that has set aside twenty-three thousand of those acres—more than a quarter of the total—as a wildlife refuge and utilizes crop rotation, cover crops, and composting. Its herd of sixteen thousand dairy cows feeds on the peels left after its potatoes have been processed into enough French fries to feed four million people. Its "closed-loop system" sounds suspiciously like a biodynamic farming system.

But this isn't biodynamic; it's realistic. It's feeding a lot of people, inexpensively. How green, by contrast, is an obscure and (arguably) costly form of viticulture that involves trucking in the manure of off-site cattle and results only in a few hundred bottles of (typically) very expensive wine? "You take your biodynamic grapes and put them in an oversized and overpriced two-kilo glass bottle and put that on the shelf? That's not a very green thing to do," grumbles one viticulturist.

Which brings us to the M-word: marketing, or "the perceived perspective of the end user or the buyer," as Patty Skinkis of Oregon State University so diplomatically puts it. Yes, biodynamic viticulture is a totally obscure niche that is known only to aficionados. But it's still plausible that some biodynamic dabblers could be dipping their toes into the vortex solely to be able to brag about it on their Web sites.

"Some of the world's most renowned wineries farm Biodynamically and many consider Biodynamics to be the 'Rolls Royce' of organic farming," Stuart Smith, vineyard manager and winemaker at the Napa Valley's Smith-Madrone winery, observes in his anti-biodynamics blog, entitled *Biodynamics is a Hoax*. "Yet, after reading Steiner, I conclude that Rudolf Steiner was a complete nutcase, a flimflam man with a tremendous imagination, a combination, if you will, of an LSD-dropping Timothy Leary with the showmanship of a P.T. Barnum."

In Oregon, disgruntled biodynamic-deniers tend to bring up the M-word in reference to Cooper Mountain Vineyards, which claims on its Web site to be "Oregon's Organic & Biodynamic Winery," in a sense co-opting the biodynamic title for itself. For a visitor arriving at Cooper Mountain for the first time, then, it's startling to be faced with the sight of McMansions tucked in amongst the vines. Is this the face of biodynamic agriculture? "That's one of the problems with certification," shrugs Demeter USA executive director Jim Fullmer. "There's nothing in the regulations that says you can't sell land off for houses, as long as you keep a 10 percent buffer for biodiversity."

(In owner Robert Gross's defense, he couldn't have known that the then-agrarian part of Beaverton where he planted vines in 1978 would have morphed into a slice of suburbia over the next three decades. He's been getting good prices for the parcels—during the research of this book, a real estate record showed a lot for sale for $375,000, with similar lots selling for similar prices over recent years—which has allowed him to purchase three additional vineyard sites, with the long-term plan of moving the winery to one of these more rural locations.)

But at least Cooper Mountain has been Demeter certified since 1999, McMansions or not. For the critics of biodynamic viticulture, the worst offenders are those who claim to be practicing the farming system but have not sought independent third-party certification through Demeter USA. As one vineyard manager fumed to me, "They're like zealots who don't even go to church!"

Demeter USA executive director Jim Fullmer agrees that this is a problem: "We actually pretty commonly send out cease-and-desist letters to whomever is representing their commercial product as biodynamic without any verification behind it," he says. And yet, the term "biodynamic" continues to be thrown around by a number of vineyard owners who have not pursued certification. Of thirty-one (at last count) Oregon vintners claiming to incorporate biodynamic practices into their farming during the writing of this book, just sixteen, or 52 percent, were either Demeter certified or in the certification process.

Similarly, in Burgundy, where some sixty domaines claimed (at last count) to practice *biodynamie*, just sixteen, or 26 percent, were Demeter certified. There's a cultural explanation for the disconnect in France, though. According to the importer Martine Saunier of Martine's Wines, Inc., the French marketplace has little awareness of *biodynamie*, because this style of horticulture is already so ingrained in French tradition. "When I was a kid, a gardener would never plant a seedling before the full moon," she recalls. "Even now, even if you don't do *biodynamie*, everybody in France looks for the full moon before they start their bottling."

Saunier grew up in Paris and Burgundy; today, her company is based in Novato, Marin County, California, and her clients are American wine merchants; and so the cover of her Autumn 2009 catalog is a wheel-shaped zodiac, a reference to the horoscopic mosaic that adorns the entrance to the cellar of the Burgundian Domaine Leroy, the jewel in Saunier's portfolio. It's a signal to clients that she carries biodynamic labels (both certified and not). "In California and on the West Coast, we have a wonderful market for biodynamic producers. But in France, there are few stores in Paris that focus on organic and biodynamic," she says. "And they're not exactly Whole Foods."

In the U.S., the biodynamic practitioners who choose not to be certified claim to follow the practices purely for the health of their vines, their property, and their laborers—*not* the marketing angle. Why, they ask, should they bother going to the trouble of filling out all that paperwork and paying all those fees to verify for the public a practice that is intensely personal?

"I've heard that argument, too. That it takes too much time in fees and paperwork," observes Patty Skinkis of OSU. "That might be a perception, but it is not reality. The paperwork isn't so hard nor the fees so costly." In fact, the paperwork isn't any more onerous than the spray schedules that all viticulturists fill out as a matter of course, and annual fees and inspection costs are comparable to those for organic certification. "I think a greater concern for them is that, if they are certified, they will lose certain tools in the toolboxes," Skinkis ventures. "Pest management, weed control, and disease management are all concerns in those years when we have rains before harvest. They don't have the flexibility to use some of those tools if they are certified. Which is why many of them are not certified."

According to vineyard manager Joel Myers, fungicides and pesticides were applied in solutions of *pounds* per acre back in the 1980s; today, thanks to advances in agricultural science, only *ounces* are used: "We can look at the active ingredient and we have reduced it between 86 and 90 percent," he marvels. "I don't want those tools taken away. By taking away those tools, you add a substantial amount of stress to the equation," he says. "And our goal is to deliver disease-free fruit to the clients. They want absolutely clean fruit." Myers has seen the devastating financial effects organic and biodynamic vineyards can have on their hardworking vinetenders. When suddenly hit by powdery mildew, a certified-organic vineyard can't resort to fungicide. The crop tainted, the buyers walk away. And a year's hard work goes down the drain.

There's a fallacy in fair-weather organics. Join an organic CSA and be prepared for the week when half the fruits and veggies in your box are rotten and inedible. The commitment to going green entails taking on some risk. The problem with grape farming is that the stakes are sky-high, because there is just one single annual harvest. If your crop rots, the year is a total loss.

"I really don't think anyone in the wine community is truly doing biodynamics, even the Demeter-certified people," reflects Todd Hamina, proprietor of Biggio Hamina Cellars in McMinnville. "Because no one works in a closed-loop system. Ehrenreid Pfeiffer stated that you can't do biodynamics in a monoculture; and yet that is what grape growing is. To do BD, you've got to have the cows on site; you need to be making the preparations yourself."*

A former winemaker for Maysara, Hamina guided Momtazi Vineyard through the Demeter certification process in 2005, six years after converting his own farm to his own form of biodynamic agriculture. Nowadays, he doesn't do much other than apply a few homeopathic teas to his vines. "It's not that I think it's super-crazy and wacky—which it is—but really, at the end of the day, you can achieve similar results without doing all that," he reflects. "I extracted as much as I could from biodynamics while minimizing the kookiness. It doesn't matter what Jupiter is doing. It does make a difference what the moon is doing, but that's not BD. That's just old-school.

"If biodynamic agriculture gets you on the path to being a better farmer, you should do it," Hamina continues. "Will BD give you a better product? Not necessarily. I'd rather get fruit from a well-managed conventional vineyard than a biodynamic vineyard that's behind the eight-ball. If you suck at canopy management, if you suck at shoot positioning, you're just pissing up a rope."

---

* It's common practice among American biodynamic practitioners to order the preps in powder form from the catalog of the Josephine Porter Institute for Applied Bio-Dynamics in Woolwine, Virginia.

Alex Sokol Blosser, like Todd Hamina, is a reformed biodynamicist. "Because we can't use chemicals, I do a lot of the stuff that biodynamics calls for. There is some voodoo stuff that I do. The compost? Voodoo. Humic acid? A little bit of voodoo. They're just leaps of faith that I know that I am making," he admits. "One thing that gets my goat is when someone says, 'Biodynamics is beyond organics.' It's like, 'Come on, Dude.' We're as into compost as the biodynamic guys are. We just don't use cow manure and we just don't do prep sprays.

"At the end of the day, it's all about farming for quality. If you farm for quality—whether you're organic, biodynamic, sustainable, or conventional—you're going to make great wine. Because we farm organically, do we make better-quality wine? No, it's because we farm for quality that we make better-quality wine. We could farm organically and pull in five tons an acre and make crap," Sokol Blosser continues. "But we don't do that. If we've got to go out a second or third time to drop fruit, we'll do it. The shoot thinning, the shoot positioning, the lateral thinning, the leaf pulling, it goes on and on. If I were to make a cheap wine, I wouldn't do any of that. I would reduce costs. I would mechanize as much as I could. But we're trying to be in the top echelon, in terms of quality, for pinot noir and pinot gris in the state. So that means, you farm for quality, and how you decide to farm is up to you."

CHAPTER TWELVE

Laly Ahnkuttie

No great thing is created suddenly, any more than a bunch
of grapes or a fig. If you tell me that you desire a fig, I
answer you, that there must be time. Let it first blossom,
then bear fruit, then ripen.

—Epictetus (A.D. c. 50–c. 138), *Discourses*

What does paradise look like? Picture a rolling green field dotted with
sheep, goats, and Highland cattle, those russet-colored shaggy
beauties with their big, brown, limpid eyes. The green pasture is
punctuated by a few white oaks; overhead is a blue sky, evocative
with billowy clouds. There is a creek lined with horsetail at the
bottom of the slope, a henhouse on a trailer parked next to the
craftsman house at the top of the hill. It's spring, so white trilliums,
wild onions, and various wildflowers are just beginning to poke out
of the ground. Native grasses, barley, wheat, and rye are sprouting
up between the rows of knobby grapevines. A light breeze stirs up
birdsong and causes the leaves to rustle in the trees.

And then, the quiet is broken by a buzzing that becomes a roar as a
four-wheeled Honda ATV comes tearing down the hill at a terrifying
clip. Just when I think I might become roadkill, the vehicle comes
screeching to a halt. Brian O'Donnell hops off and ambles toward
me. Bearded, bespectacled, soft-spoken, and thoughtful, O'Donnell
could pass as a New England college professor. That anomalous
performance on the ATV was a head-scratcher, the only hint he offers
that, living here on this piece of paradise, he feels like a kid in a
candy store.

Along with his wife, Jill, and daughter, Riona, Brian lives on
seventy acres just outside of the picturesque village of Carlton.
They call their estate Belle Pente, French for "beautiful slope." The
property includes sixteen acres of vineyard, planted on gently rising
hillsides that range from two hundred forty to five hundred feet,
with mostly southern exposures.

Here in the United States, it's tempting to deem a picturesque
vineyard property stocked with a menagerie of animals a
"gentleman's farm." But in other parts of the world, farmsteads that

integrate a wide variety of livestock and plant species—fruit and olive trees, a vegetable garden, a cow, a horse, a pig, chickens, cats, dogs, a few rows of grapevines—are the sole source of subsistence for their owners. Whether it's in an Italian hilltown or a village in the shadow of the Andes, this is a long-established, but fast-disappearing, model of sustainable agriculture.

Which begs the question: Is such a thing a farm or a vineyard? Which begs another question: Why do we consider the two to be separate entities?

At Belle Pente, the vines finance the animals, and the animals nourish and manage the vines. Most of the year, sheep graze in the meadow, keeping the grass down and preventing it from going to seed; the free-range lambs are sold for meat. But in late winter and early spring, the wooly quadrupeds amble down the vine rows, gobbling up the grasses and weeds that sprout up between the vine trunks. A contingent of chickens follows, feasting on the pernicious cutworms that threaten to damage tender buds and shoots.

A couple of times a year, Brian O'Donnell cleans out his barns and combines straw, animal manure, and winery waste (grape skins, stems, and seeds) to make a compost pile. He lets each pile sit a year and decompose; then he spreads it over his vineyard. He also makes biodynamic barrel compost, mixing his cows' manure with eggshells from his henhouse.

Lately, the O'Donnells have been using whey from Tillamook Dairy as a fungicide and have found it to be effective. Their Highland cattle are a meat breed, a bit too wild to be handled and milked, so the O'Donnells are considering purchasing an Irish Kerry cow or two, for milk, cheese, and whey production.

Brian and Jill originally planted the vineyard in 1994; they began farming organically in 2000, when the birth of their daughter caused them to question why they were dispersing chemicals so close to their home. In 2002, when they decided to pursue organic certification, they attended a meeting at Cooper Mountain Vineyards with representatives from Oregon Tilth, the organic certification organization, and Demeter. "Oregon Tilth had sort of a Ten Commandments view of the world: thou shalt not do this, thou shalt not do that. Whereas the Demeter approach was, 'Here are things you can do to create a healthy environment and increase the natural disease resistance of the plants,'" O'Donnell recalls. "It was more of a proactive thing, rather than a reactive thing. So that kind of intrigued me."

Since then, Brian O'Donnell has completed most or all of the requirements for biodynamic certification each year. And he has seen results: In 2007, he implemented the full program "and we were probably three to five days ahead of our neighbors in maturity. It allowed us to pick early, a few days before everybody else, and get most of our stuff in before the rains." Another year, he applied the 500 preparation just prior to a cold snap: "There was frost all around the perimeter, but no frost on the vines. It was pretty amazing."

O'Donnell tries to be mindful of the lunar calendar when planting, pruning, and harvesting. He sprays the biodynamic preparations, when time allows. "You go out in the vineyard after you've done an application and look at the angle and color of the leaves. The plant just looks like it's vibrating, everything is kicking on all cylinders," he says. "The flip side is, it's another pass through the vineyard, which uses more fossil fuels."

Why isn't Belle Pente certified biodynamic? O'Donnell shrugs. "Our goal is to make great wines in an environmentally responsible way, and we're not hung up on the whole idea that we have to make organic or biodynamic wines." He thinks of the times he has had to acidify, or use a commercial yeast to restart a stopped fermentation: "I'm not willing to compromise quality for the sake of philosophy."

For the O'Donnells, biodiversity is important. A self-sustaining farm that integrates livestock is important. Reducing the use of fossil fuels is important. Farming naturally is important. Growing grapevines without irrigation is important. A seal on a wine label is not important. Demeter USA executive director Jim Fullmer wistfully calls Belle Pente "a beautiful farm" and "a biodynamic farmer's dream."

"Goethe was trying to create this morphological consciousness. He was able to look at a sprout and tell you what the flower would look like," a biodynamic consultant once told me. "Knowing the forms is a way in. It's a way of putting your hand on the door and getting in, in a more sacred way. The language of form is what I feel Rudolf Steiner was asking us to train ourselves in."

Belle Pente isn't just a biodynamic vision of beauty. It's everyone's vision, really: the idyllic sloping meadow dotted with animals; the neat rows of grapevines. It's the Platonic ideal, the Goethean final form, of what a farm should look like.

In Ghana, the word *sankofa* means "go back and retrieve"; it refers to the notion of learning from history.

The agricultural movement that encompasses organic and biodynamic farming, locavorism, the farmers market, and the CSA is one of *sankofa*. A year-round schedule of seasonal crops and market days, a farming style that requires plenty of sweat and planning: until recently, many Americans were only aware of these folkways from foreign travel and historic literature.

For wine lovers, certain aromas and flavors stir remembrances of things past. Very occasionally, we'll come across a wine today that reminds us of one we tasted three decades ago. Or a fifty-year-old red that is just beginning to brown around the edges, still a marvel to behold and savor. The best vintners are constantly looking back, trying to decipher what combination of farming, harvesting, and vinification techniques created those classics so that they can attempt to replicate the process today.

Brad Ford is such a vintner. When he speaks of future plans for his family's Illahe Vineyards and Winery in Dallas, Oregon, there is no talk of purchasing modern equipment or of streamlining his process. Instead, he mentions his dream of fermenting pinot gris in buried terra-cotta pots, as is the tradition in the central-European nation of Georgia and in parts of Italy. He's looking forward to the day when he can afford to construct underground caves for cellaring.

Brad's father, Lowell, began growing grapes and making wine in 1983, but didn't purchase the Illahe property until 1999, when the cattle rancher next door put the eighty-acre parcel up for sale. Today, Brad helps their neighbor with cattle drives; they trade wine for beef and sell their siegerrebe grapes to the rancher's wife.

Brad Ford's biggest equipment purchase of the past couple of years has been an investment in two Percherons, Doc and Bea. Smart and agile, these stately French-bred draft animals can pull heavy loads and were invaluable to foresters and farmers prior to the invention of the mechanical engine. "I thought we could reduce the tractor runs, reduce the impaction of the soil, do something that is romantic and beautiful and slow down the process," says Ford. "The first day you turn on a new tractor is the best day you're going to have with it. But a horse gets smarter and smarter every day you work with him, and you get better and better at working with him. A day with a horse is far more satisfying than a day spent on a tractor."

Draft horses have become emblems of the biodynamic farm; top biodynamic vineyards in other regions throughout the world use them. Notably, French native Christophe Baron of the Cayuse label has named one of his Walla Walla vineyards (which is in fact located just over the Washington border, in eastern Oregon) "Horsepower," in honor of the draft horses that work it.

Brad Ford isn't farming biodynamically. But he is going out into his vineyard every day and—instead of starting up an engine and turning a steering wheel—holding a harness and muttering commands like "haw" and "gee." Doc and Bea haul grapes to the winery at harvest time and mulch and mow the vine rows. Each time they master a new task, they get closer to putting the Fords' tractor out of business.

In the winery, Brad Ford and his co-winemaker, Michael Lundeen (who has worked for the traditional Barolo and Barbaresco producer Castello di Verduno and cites Brian O'Donnell's Belle Pente winery as a personal inspiration), punch down grapes by foot as much as possible and use a wooden basket press. They're saving their pennies for another wood fermentation tank. "Sure, it's a lot of work, but I want to see what kind of wine I can make without mechanical or chemical intervention. I'm looking backward, in history, to learn how to make my wine," says Ford. "The philosophy is anticorporate soda-pop. We want *terroir*. We want this wine to taste like this hill. That's why our name is Illahe. It means 'soil' or 'land' in the Chinook jargon, the pidgin language tribes used to trade with one another around here."

Another good name for this estate might be "Laly Ahnkuttie." Back when the people in these parts spoke Chinook, this term came up often. It means "long ago" or "once upon a time."

Clare Carver and Brian Marcy have a hundred laying hens. They have fifty free-range meat chickens. They have five Irish Dexter cattle, two Pigmy-Nigerian Dwarf goats, two Nubian goats, one guard llama, three Duroc-Berkshire pigs, two Haflinger draft horses, one overfed swarm of approximately ten thousand bees, one hound dog, and one farm cat.

Laly Ahnkuttie

Clare Carver and Brian Marcy have no vines. Yet.

They have set aside twenty acres of well-drained, south-facing hillside; they're pulling out stumps and the goats are clearing the blackberries. They'll plant pinot noir, chardonnay, riesling, and even syrah. And then, like the O'Donnells do at Belle Pente, they'll send the goats and chickens down the rows to weed and fertilize. And like the Fords do at Illahe, they'll send their draft horses through to spread manure.

The focal point of their Gaston property, Big Table Farm, is a purplish-pink Victorian built in 1890 by Joseph Williams, for whom the surrounding Williams Canyon is named. On the porch is a box full of fuzzy yellow chicks, warming under a heat lamp. In the yard, between some twisted old pear and apple trees, is a chicken "bus": a covered coop on wheels, full of older chicks that are starting to show their black and white speckles. Mature chickens wander freely around the house, pecking at the dirt.

Inside, a dog sleeps on a blanket in front of the stove in the living-cum-dining room, the centerpiece of which is the eponymous row of "big" tables draped with cheery cloths, printed with bright yellow borders and pale-red flowers and pomegranates. A motley assortment of vintage chairs and benches pushed underneath could seat twenty guests; the couple regularly hosts large, raucous dinners for Portland chefs, local farmers, and frequent customers.

The walls are a rich cream, the stairs a saturated red. A guitar is propped in one corner of the room, an African spear in another. In a light-filled adjoining office are an easel and an antique drafting table, its narrow drawers overflowing with half-used tubes of paint. Carver's paintings and sketches hang on the walls in small frames: tractors, chickens, silverware, barns.

The kitchen counter is cluttered with Mason jars filled with preserves, pickles, dried mushrooms, macerating limoncello, and homemade honey. Brian cures his own prosciutto, pancetta, and guanciale and smokes his own ham and bacon; Clare grows and either serves or preserves fresh produce. "I can like crazy in the fall. Our electric bill pretty much doubled last September and October. I was running the stove nonstop," she says cheerily. "I grow thirty-five tomato plants and put up seven gallons of tomatoes each year. Anything else you can name, I'm growing it: broccoli, Brussels sprouts, strawberries, tomatoes, beets, onions, chard, spinach, fava beans, peas."

It's a wet early-May day, the sound of the rain hitting the roof is broken periodically by the crackling fire and the chickens clucking outside. Brian and Clare have finished their morning chores and sit by the wood stove, eating eggs with yolks as yellow as small suns. Clare has reddish-auburn hair and a wide, toothy smile; she's the sort of person who greets you with a big hug on your first meeting. Brian is more soft-spoken, with twinkling blue eyes and a boyish face that belies his determination.

Throughout the property's seventy hillocky acres, Brian's work is evident: Since the couple moved here from Napa in the fall of 2006, he has rebuilt the barn using salvaged timber. From scrap metal that was lying around the property when they bought it, he has fashioned the mesh-bottomed chicken bus, the sides of which are covered with wooden flaps that fold up to reveal hens sitting on eggs. Also, there's the "Pig-ebago," and more mobile homes for goats, cows, and horses. The point of these wheel-mounted sheds is to provide shade, protection from the elements, and rain-filled water troughs to the animals, wherever they are grazing.

Marcy and Carver farm using the same method extolled by Virginian Joel Salatin in Michael Pollan's 2006 book, *The Omnivore's Dilemma*. At his family farm, Polyface, Salatin introduces Pollan and his readers to a self-sustaining style of free-range farming that mimics nature. As herds of American bison once moved in tightly massed groups as they covered the wild open range, so as to be protected from predators by their numbers, the high-intensity grazing technique sequesters animals in packs, over relatively small parcels of land. At Big Table Farm, this is achieved by setting up white nylon solar-powered electric fences in one small section of field at a time. At each stop, the cows eat grass, pull weeds, fertilize the soil with their manure and work it with their hooves. Then they move on, allowing that section of soil to regenerate.

"It's basically using animals to manage the grass," Marcy explains. "The crop is grass. That's a mentality that would make us more productive and change our food system. They're finding in the Midwest right now that it's actually more profitable to graze areas where they are farming row crops, like corn and soybeans, that are destined to become cattle feed. You don't have to drive a tractor, you don't have to buy fuel. It's just letting animals do their own work to feed themselves."

"We've spent a lot of time and energy reading about intensive grazing systems and pasture management and building top soil and carbon sequestration," adds Carver.

"The typical concept of grazing in the United States is, 'Fence the whole place and throw your animals in it and let 'em eat whatever they want,'" Marcy continues. "But studies have shown that even if you divide your field up into just four cells, it's more effective than letting them roam all over a big pasture. It really enriches the soil and creates a diversity of flora and fauna."

In the midst of a discussion like this, it's easy to forget that farming is just one of many occupations for this couple. Carver is a painter and graphic designer who has created or redesigned labels for top wineries such as Grace Family Vineyards, Miner, Thomas Brown, Beaux Frères, Domaine Drouhin Oregon, Matello, and ADEA. Marcy is an accomplished winemaker whose resume includes California stints with Marcassin and Neyers Vineyards. Here in Oregon, he is the winemaker for Coelho and consults for other wineries.

Marcy used to be the oenologist for Resonance Vineyard as well, until owner Kevin Chambers gave up winemaking. Now Marcy and Carver buy Chambers' biodynamically grown Resonance fruit and vinify it for their own Big Table Farm label. They are, at present, animal farmers who happen to make wine from biodynamic grapes. When Chambers invited the couple to join his occasional biodynamic discussion group, Marcy and Carver demurred. On top of all their animal husbandry and fence moving and vegetable growing and cooking and painting and designing and winemaking, they just didn't have time. They're intrigued, though. They've been giving cow manure and eggshells to their friend Dan Rinke at Johan Vineyards as well as other BD vineyard managers, such as Stirling Fox.

"We're just getting started," says Marcy. "I don't think we even know enough yet to apply something like biodynamics. We're just getting our wheels under us. It's almost like a religion. How do you believe in it when you are still learning about the basics?"

Carver, a lapsed Catholic, admits that she was initially put off by Steiner's religious ramblings, but she's still reading, discussing, testing. She and Marcy are interested in the possibility that the phases of the moon might affect vine growth and wine quality. "When we were moving up here from Napa, I wanted to bring some of my roses. So I did a thirty-day trial and did cuttings every day and handled them all exactly the same way and put them in bags with dirt and marked each one. And it was right in keeping with the curve that they claim in the books," says Carver. "It was my own little moon-cycle test. It was really simple. I still have the bushes in my garden to prove it."

"The thing about biodynamics is that it's based on thousands of years of observation," says Marcy. "The way I learned to make wine was without adding anything, just letting the wine make itself. To me, the biodynamic preparations are in some ways similar to making wine in that, if you create the right environment, the wine is going to make itself because everything is already there. With the preparations that you spray, you are helping create an environment for beneficial organisms to flourish."

"But a lot of the vineyards, to get certification, they're buying their preparations. And I have to say that I don't agree with that," says Carver. "This might make people mad, but you know that saying, 'There's no better fertilizer than the farmer's footsteps'? The point of biodynamics in terms of the preparations, it seems to me, is that you have to know what yarrow root is, you have to know where the northwest corner of your property is, you have to know what the moon cycle is, you have to be paying attention to all of these things. What if there's some microorganism that came from your particular property that could actually really be beneficial in your compost heap that isn't there if you buy your preparations from Virginia? I guess we're just getting there more slowly. I think there are pieces of it that we do hope to integrate. But do we want to call ourselves 'biodynamic'? I don't know ..."

Marcy looks at his wife fondly, shaking his head.

"Clare," he says, "some people just farm."

Notes on sources

PREFACE

"The access road was hemmed in": Katherine Cole, "Advocates of 'Ultra-Organic' Farming Say It Creates Better Wines." *The Oregonian*, June 8, 2003, p. L7.

INTRODUCTION

epigraph: Cole Porter, "You Do Something to Me," from *Fifty Million Frenchmen*, 1929. Reprinted with permission from Alfred Music Publishing.

"the season when the Earth is most inwardly alive": Rudolf Steiner, *Spiritual Foundations for the Renewal of Agriculture*, trans. Catherine E. Creeger and Malcolm Gardner, ed. Malcolm Gardner (Kimberton, Pennsylvania: Bio-Dynamic Farming and Gardening Association, Inc., 1993), p. 73.

"with quartz that has been ground": Ibid., p. 74.

"enliven the soil": Ibid.

"For this observer, biodynamic processes": Matt Kramer, *Matt Kramer on Wine: A Matchless Collection of Columns, Essays, and Observations by America's Most Original and Lucid Wine Writer* (New York: Sterling Epicure, 2010), p. 117.

CHAPTER ONE

epigraph: *The Hummingbird's Daughter*, by Luis Alberto Urrea (New York: Little Brown and Company, 2005(, p. 97, 99.

Persia had a winemaking tradition: Gregory McNamee, *Movable Feasts: The History, Science, and Lore of Food* (Westport: Praeger Publishers, 2007), p. 92.

"resist the so-called menace of communist encroachment": Ahmad Ashraf, "From the White Revolution to the Islamic Revolution," in *Iran after the Revolution: Crisis of an Islamic State*, ed. Saeed Rahnema and Sohrab Behdad (London: I.B. Tauris & Co., 1996), p. 22.

Archaeologists place the first use: Peter Bellwood, *First Farmers: The Origins of Agricultural Societies* (Malden, MA: Blackwell Publishing, 2005), p. 51.

As far back as the sixth millennium: Robert Chadwick, *First Civilizations: Ancient Mesopotamia and Ancient Egypt* (London: Equinox, 2005), p. 112.

The Greeks of antiquity: Richard G. Olson, *Technology and Science in Ancient Civilizations* (Santa Barbara: Greenwood, 2010), p. 96.

gave us the poet Hesiod: Monty Waldin, *Biodynamic Wines* (London: Mitchell Beasley, 2004), p. ix.

"Do not rise or sit or eat": Steven J. Williams, *The Secret of Secrets: The Scholarly Career of a Pseudo-Aristotelian Text in the Latin Middle Ages* (Ann Arbor: University of Michigan Press, 2003), p. 12, 16.

Continuing a tradition seven millennia old: Mauro Ambrosoli, *The Wild and the Sown: Botany and Agriculture in Western Europe, 1350-1850*, trans. Mary McCann Salvatorelli (Cambridge: Cambridge University Press, 1997), p. 32, 77.

"Crop failures might occur as often": Ken Albala, *Food in Early Modern Europe* (Westport, Connecticut: Greenwood Press, 2003), p. 8.

"*Sow peason and beans in the wane*": Thomas Tusser, *Some of the Five Hundred Points of Good Husbandry,* ed. H.M.W. (Oxford: John Henry Parker, reprinted 1847), p. 66.

"*in the best of situations*": Albala, p. 13.

"*The appearance of a few drops*": Tom Standage, *An Edible History of Humanity* (New York: Walker & Company, 2009), p. 199.

"*The scheme for the constant improvement*": Ambrosoli, p. 396.

"*A careful study of the condition*": Frank D. Gardner, *Traditional American Farming Techniques* (Guilford, Connecticut: The Lyons Press, 2001; original title *Successful Farming,* published by L.T. Myers in 1916), p. 72.

According to the National Center for Farmworker Health: "Migrant and Seasonal Farmworker Demographics" (Buda, Texas: National Center for Farmworker Health, Inc., 2009), p. 1.

"*...The fact remains that the man who made possible a dramatic expansion of the food supply*": Standage, p. 212.

CHAPTER TWO

The only remotely meaty tome: Gary Lachman, *Rudolf Steiner: An Introduction to His Life and Work* (New York: Jeremy P. Tarcher/Penguin, 2007).

At the same time, young Rudolf: Steiner, *Autobiography: Chapters in the Course of My Life: 1861-1907* (Great Barrington, MA: SteinerBooks), p. 5.

Footnote #37: Rudolf Steiner, *The Influences of Lucifer & Ahriman: Five Lectures by Rudolf Steiner* (Hudson, NY: Anthroposophic Press, 1993), p. 9.

References to "devil" and "Satan" in Footnote #37: Rudolf Steiner, ed. Robert A. McDermott, *The Essential Steiner: Basic Writings of Rudolf Steiner* (Edinburgh: Floris Books, 1996), p. 374.

"*Although decidedly Germanic*": Lachman, p. 6.

according to a 2008 study: Nicholas Epley et al., "Creating Social Connection Through Inferential Reproduction: Loneliness and Perceived Agency in Gadgets, Gods, and Greyhounds," *Psychological Science* 19:2 (February 2008), pp. 114-20.

There must have been something in the water: As discussed by author Thomas G. West in *Thinking Like Einstein: Returning to Our Visual Roots with the Emerging Revolution in Computer Information Visualization* (Amherst, NY: Prometheus Books, 2004) and *In the Mind's Eye: Visual Thinkers, Gifted People With Dyslexia and Other Learning Difficulties, Computer Images and the Ironies of Creativity* (Amherst: Prometheus Books, 1997).

Unlike Tesla, Steiner didn't have visions: Lachman, p. 149.

"*Of all the spiritual thinkers of the twentieth century*": Ibid., p. 19.

Nietzsche's soul hovering above his head: Rudolf Steiner, *Autobiography,* p. 130.

Nietzsche himself would have found ... nauseating: Lachman, p. 84.

Footnote 47: Richard G. Olson, *Technology and Science in Ancient Civilizations* (Santa Barbara: Greenwood, 2010), p. 97.

Swedenborg, who "used ... sexual energy to see spiritual worlds": Rudolf Steiner, "Sexuality and Modern Clairvoyance: Freudian Psychoanalysis and Swedenborg as a Seer (Dornach, September 14, 1915)," in *Community Life, Inner Development, Sexuality and the Spiritual Teacher* (Hudson, NY: Anthroposophic Press), p. 78.

magical powers and vehicles that hovered in mid-air: Rudolf Steiner, with Andrew Welburn: *Atlantis: The Fate of a Lost Land and Its Secret Knowledge,* trans. C. von Arnim (London: Rudolf Steiner Press, 2007).

Sources

The theosophical mystics were: Steiner, *Spiritual Foundations for the Renewal of Agriculture,* p. 165.

He designed and oversaw the construction: Rudolf Steiner, *Autobiography,* p. 337.

In 1919, as Germany considered: Ralph Courtney, "Futurum's Promises: New Plan 'to Regenerate Industrial System' Attracts Attention in Europe," *The New York Times* (July 17, 1921), p. 81. Johannes Hemleben, *Rudolf Steiner: An Illustrated Biography,* trans. Leo Twyman (London: Sophia Books, 2000), p. 168.

"gnomes, undines, sylphs": Rudolf Steiner, *Agriculture: An Introductory Reader* (Forest Row, UK: Sophia Books, 2003), p. 158.

They're also eerily prescient: Rudolf Steiner, *From Comets to Cocaine … Answers to Questions,* trans. Matthew Barton (London: Rudolf Steiner Press, 2000), p. 228.

"All this takes a certain amount": Steiner, *Spiritual Foundations for the Renewal of Agriculture,* pp. 101-102.

This streak of achievement: Lachman, p. 220.

Many critics of the Waldorf pedagogy: David Ruenzel, "The Spirit of Waldorf Education," *Education Week* 20:41 (June 20, 2001), pp. 38-45.

"dream imagery": Terry Gross, "James Cameron: Pushing the Limits of Imagination," *Fresh Air,* February 18, 2010.

CHAPTER THREE

epigraph: Steiner, *Spiritual Foundations for the Renewal of Agriculture,* p. 134.

Studying tables of climate conditions: The Steeles' data source: John Gladstones, *Viticulture and Environment* (Broadview, Australia: Winetitles, 1992).

The largest concentration of intact watersheds: Information courtesy of Klamath-Siskiyou Wildlands Center, www.kswild.org.

"regenerative rather than degenerative": "Guidelines and Standards for the Farmer for Demeter Biodynamic®, In-Conversion-to-Demeter Biodynamic® and Aurora Certified Organic®" (Philomath, OR: Demeter Association, Inc., 2005, updated 2009), p. 5.

habitat for beneficial plants, insects: A study led by three Washington State University entomologists concluded that the largest and healthiest potato plants grow in plots not treated with insecticides, and where native beneficial insect populations flourish. Their findings were published recently in the journal *Nature:* David W. Crowder, et al., "Organic Agriculture Promotes Evenness and Natural Pest Control," *Nature* 466 (July 2010), pp. 109-12.

"A cow has horns in order to send": Steiner, *Spiritual Foundations for the Renewal of Agriculture,* p. 71.

In the Gatha Ahunavaiti verses: Martin Haug, *Essays on the Sacred Language, Writings and Religion of the Parsis* (London: Routledge, reprinted 2002; originally printed 1884 as part of Trübner's Oriental Series), pp. 147-48.

"Why the cow was selected": Om Prakash Misra, *Economic Thought of Ghandi and Nehru: A Comparative Analysis* (New Delhi: M D Publications, 1995), pp. 34-35.

"with the preparations representing": Waldin, pp. 7-8.

"you have to start stirring": Steiner, *Spiritual Foundations for the Renewal of Agriculture,* p. 73.

In fluid dynamics: Malcolm W. Browne, "Deadly Maelstrom's Secrets Unveiled," The New York Times (September 2, 1997), p. C1.

"Then you reverse direction": Steiner, Spiritual Foundations for the Renewal of Agriculture, p. 73.

"Just 'do this and then that.'": Ehrenfried E. Pfeiffer, *New Directions in Agriculture*, excerpted as Appendix C to *Spiritual Foundations for the Renewal of Agriculture*, pp. 258-59.

although a few scientific studies have found: In 2001, two chemists at the Kwangju Institute of Science and Technology in South Korea made the "unusual observation" that when they diluted a carbon solution, the carbon molecules clumped together instead of drifting farther apart. The more water they added to the solution, the larger the carbon clusters. Shashadhar Samal and Kurt E. Geckeler, "Unexpected Solute Aggregation in Water on Dilution," *Chemical Communications* 21 (2001), pp. 2224-25.

These familiar-smelling liquids are a reminder: Steiner, *Spiritual Foundations for the Renewal of Agriculture*, p. 100: "Since plants are simpler than animals or humans, healing can take place on a more general level. With plants you can use a kind of universal remedy."

"It's important to note that we never get": Steiner, *Spiritual Foundations for the Renewal of Agriculture*, p. 96.

Then there's the certification process: "Fee Schedule" (Philomath, OR: Demeter Association, Inc., 2009), p. 4.

"Observation of the Biodynamic calendar": "Demeter USA Wine Making Standards" (Philomath, OR: Demeter Association, Inc., July 2009), p. 2.

CHAPTER FOUR

epigraph: Leonard Mlodinow, "A Hint of Hype, A Taste of Illusion," *The Wall Street Journal Weekend Edition* (November 14-15, 2009), p. W2.

"The equations to calculate the tidal forces": Joe Eskanazi, "Voodoo on the Vine," *SF Weekly* (November 19, 2008), accessed online at http://www.sfweekly.com/2008-11-19/restaurants/voodoo-on-the-vine/1/.

"We can see why it is difficult": Jamie Goode, *The Science of Wine: From Vine to Glass* (Berkeley/London: University of California Press/Mitchell Beazley, 2005), p. 68.

In another experiment: Jennifer R. Reeve et al., "Influence of Biodynamic Preparations on Compost Development and Resultant Compost Extracts on Wheat Seedling Growth," *Bioresource Technology* 101 (July 2010), pp. 5658-66.

"[T]he study consisted of two treatments": Jennifer R. Reeve et al., "Soil and Winegrape Quality in Biodynamically and Organically Managed Vineyards," *American Journal of Enology and Viticulture* 56 (December 2005), pp. 367-76.

"there is little evidence": Reeve et al., p. 374.

When he read a book called The Field: Lynne McTaggart, *The Field: The Quest for the Secret Force of the Universe* (New York: HarperCollins, 2002).

a go-to-book for BD practitioners: *The Field* is cited as a source, for example, by Richard Thornton Smith in his book *Cosmos, Earth and Nutrition: The Biodynamic Approach to Agriculture* (Forest Row, UK: Sophia Books, 2009).

Popp's research has shown: McTaggart, pp. 42-51.

Sources

In 1994, *a researcher in Switzerland*: Rogier van Bakel, "Mind over Matter," *Wired* (April 1995), accessed online at http://www.wired.com/wired/archive/3.04/pear.html.

"*Quantum mechanics*": Ian McEwan, *Solar* (New York: Nan A. Talese/Doubleday, 2010), p. 19.

"*Quantum theory is invoked*": Robert Park, *Voodoo Science: The Road from Foolishness to Fraud* (Oxford: Oxford University Press, 2000), p. 208.

"*Quantum theory has unfortunately*": Robert Lanza with Bob Berman, *Biocentrism: How Life and Consciousness are the Keys to Understanding the True Nature of the Universe* (Dallas: BenBella Books, 2009), pp. 61-62. Incidentally, Lanza's book offers insights that might interest anthroposophists, including a mention that the Star of Bethlehem story "was astrological in origin" (p. 154).

"*The world of wine exists in non-Euclidean space*": from Randall Grahm's blog "Been Doon So Long" (http://www.beendoonsolong.com/blog) "On a Mission: The Germ of an Idea," November 10, 2010)

a quote from physicist Max Planck: Nicolas Joly, *BioDynamic Wine, Demystified* (San Francisco: Wine Appreciation Guild, 2008), p. 1.

"*a remarkable intuition about the direction*": Lachman, p. 257.

"*people farmed this way for thousands of years*": New York: Riverhead Books, 2010, from the introduction (accessed via Kindle).

The idea is that this energy will influence: Bonny Doon Vineyard's Randall Grahm on the old-world version of the cosmic pipe: "Philippe Viret ... practices something called 'cosmoculture,' and ... may be making some of the most distinctive wines in the world. Philippe uses no fertilizers or other treatments in his vineyard, neither sulfure nor Bordeaux mix or even compost teas. He employs ... 'cosmic pipes'... not something to be smoked, but ... stone menhirs strategically situated throughout his vineyard, to better balance the energetic field of his domaine." Randall Grahm, *Been Doon So Long: A Randall Grahm Vinthology* (Berkeley: University of California Press, 2009), p. 299.

CHAPTER FIVE

"*In no other time would a highly regarded*": Adam Gopnik, "NO RULES! Is Le Fooding, the French culinary movement, more than a feeling?" *The New Yorker* (April 5, 2010), p. 40.

"*le vin bio n'existe pas!*": Philippe Bidalon, "Michel Bettane: '*Le vin bio n'existe pas!*' (*L'Express*, March 9, 2009), Styles: "[L]e vin bio n'existe pas! Le mot «bio» ne peut en effet, dans l'état actuel de notre législation, s'appliquer qu'au raisin ... Ceux qui tiennent à cette dénomination s'acharnent à accuser les autres types de vins de trahir leur terroir et même d'être dangereux pour la santé ... La dégustation comparative démontre, en revanche, que des vins reconnus par tous comme exprimant remarquablement leur origine sont régulièrement produits à partir de raisins non bio...." Rough translation: "[O]rganic wine does not exist! The word 'organic' may in fact, in the present state of our legislation, apply only to grapes ... Those who hold this designation are bent on accusing other types of wines of betraying the local *terroir* and even of being hazardous to the health ... A comparative tasting showed, however, that wines recognized by everyone as remarkably expressive of *terroir* are regularly produced from non-organic grapes ...."

"The natural-wine movement has been sweeping": Dave McIntyre, "Natural Isn't Perfect, Naturally," *The Washington Post* (April 14, 2010), p. E5.

"free of chemical contamination": Demeter Biodynamic Guidelines and Standards, p. 13.

They might have flipped through (note): Douglas Brenner, "Soul Man: What Didn't Rudolf Steiner Do?" *T: The New York Times Style Magazine* (April 11, 2010), p. 68.

And they shop at the fiercely independent: As chronicled recently by Anthony Rose, "Declaration of Independents," *Decanter* (April, 2010), p. 64; and by Eric Asimov, "Good Wines and No Attitude. Yes, It's True," *The New York Times* (March 24, 2010), p. D1.

Steiner associate Ehrenfried Pfeiffer: Ehrenfried Pfeiffer, *Sensitive Crystallization Processes: A Demonstration of the Formative Forces* (Spring Valley, NY: Anthroposophic Press, 1975), p. 3.

"It is very variable": To see a series of sensitive crystallizations by Bonny Doon Vineyard's director of viticulture, Philippe Coderey, visit http://www.biodynamics.com/sensitive-crystallizations-coderey

"a methodology of arriving": Anthony Rose, "Strange Fruit," *The Independent Magazine* (September 12, 2009), p. 59.

"Mercury seemingly very retrograde": http://twitter.com/RandallGrahm, April 28, 2010, 11:03am PST. Number of followers as of May 5, 2010.

And when supermarkets and bottle shops: "Everybody's Talking About … Tasting in Tune with the Moon" *Decanter* (July 2009), p. 8. Also reported in mainstream British media sources, such as BBC News, *The Guardian* and *The Telegraph*. Also: Linda Rodriguez McRobbie, "Drinking with the Stars," *American Way* (March 15, 2010), p. 24.

Articles on the subject inevitably: Ben Giliberti, "Is it Voodoo, or Old-Fashioned Passion?" *The Washington Post* (February 21, 2007), p. F2. A month after I submitted the working title for this book—*The Voodoo Vintners*—an article appeared in *SF Weekly* entitled "Voodoo on the Vine."

"fundamental principle is that everything": John Burnett, "Quake Takes Its Toll on Haiti's Burial Rites," *All Things Considered* (February 3, 2010).

To the Haitians, this spiritual dimension: Barbara Bradley Hagerty, "Voodoo Brings Solace to Grieving Haitians," *All Things Considered* (January 20, 2010). As explained by "Elizabeth McAlister, a Voodoo expert at Wesleyan University."

They're eccentrics like Alois Lageder: Beverly Blanning, "Take It Easy: With a Near-Perfect Climate for Grape Growing, Winemakers in Alto Adige Have Little Reason to Worry in a Region with a Relaxed Pace of Life," *Decanter* (2010 Italy Issue), pp. 33-34.

"For lunar, read loony": Anthony Rose, "Strange Fruit," *The Independent Magazine* (September 12, 2009), p. 59.

CHAPTER SIX

epigraph: (New York: Farrar, Straus and Giroux), pp. 88-89.

One afternoon in late July 2001: My gratitude to Whitney Gauger, executive director of the International Pinot Noir Celebration, for providing details regarding this event.

Sources

"La Tigresse": Natalie MacLean, *Red, White and Drunk All Over: A Wine-Soaked Journey from Grape to Glass* (New York: Bloomsbury USA, 2006), p. 27.

The audience sniffed and sipped: Thanks to Matt Kramer, Martine Saunier, and Laurent Montalieu for their recollections of this event.

Maison Louis Jadot: BurgundyToday.com, "Biodynamic Wines," http://www.burgundytoday.com/gourmet-traveller/biodynamic-wines.htm (accessed January 13, 2010).

With their ample cellars and knowledgeable labor force: The Oxford Companion to Wine, ed. Jancis Robinson (Oxford: Oxford University Press, 1999), p. 115.

According to biodynamic wine author: Waldin, p. 93.

"In New York pregnancy is a ward": Adam Gopnik, *Paris to the Moon* (New York: Random House, 2000), p. 301.

As Wright points out: Wright cites Allen Meadows, author of the online wine journal Burghound.com, as the source of this idea.

CHAPTER SEVEN

epigraph: "Understanding Oregon Pinot Noir Country" (October, 2009), p. 44.

Widely regarded as the inventor: *Routledge Encyclopedia of Philosophy: Nihilism to Quantum Mechanics, Volume 7*, ed. Edward Craig (London: Taylor & Francis, 1998).

"I think that they believe in all the things": Charles Heying, *Brew to Bikes: Portland's Artisan Economy* (Portland: Ooligan Press, 2010).

"Come visit us again and again": Brent Walth, "Blazing Trails in the 1970s," from *An Oregon Century: 100 Years of Oregon in Words and Pictures*, published on the Web site of *The Oregonian*, http://www.oregonlive.com/century/1970_intro.html

CHAPTER EIGHT

"One reason that biodynamics has caught on": Douglass Smith and Jesús Barquín, "Biodynamics in the Wine Bottle," *Skeptical Inquirer* 31:6 (November-December 2007).

In 1905 John Forbis, a corporate attorney: Oregon State Archives, Governor Neil Goldschmidt's Administration, Box 56, Item #45.

CHAPTER NINE

"More and more vineyards have dramatically": Robert M. Parker, Jr., *Parker's Wine Buyer's Guide* (New York: Simon & Schuster), p. xx.

"Donna Morris and Bill Sweat were in finance": Matt Villano, "How to Walk a Mile in Your Dream Career," *The New York Times* (February 20, 2007), p. H6.

A 2007 Rolling Stone article: David Fricke, "The Police: A Fragile Truce," *Rolling Stone* (June 28, 2007), p. 40.

"Some of the world's most renowned wineries": From Stuart Smith's blog "Biodynamics is a Hoax" (http://biodynamicshoax.wordpress.com/, September 22, 2010, posting by commenter "Steve")

CHAPTER TEN

epigraph: Rudolf Steiner, trans. D. S. Osmond, *The Influences of Lucifer & Ahriman: Five Lectures by Rudolf Steiner* (Great Barrington, MA: SteinerBooks, 1993), p. 83.

Today the farm is gone: Maura J. Casey, "The Little Desert that Grew in Maine," *The New York Times* (September 22, 2006), p. F7.

"Soil erosion was a major reason": (Seattle: University of Washington Press, 2009)

There are sixteen million farm acres: From the National Agricultural Statistics Service "2009 Oregon Vineyard and Winery Report."

"It's estimated that 30 percent of the carbon": For more on this, check out the thirty-minute film "Big River," made by Ian Cheney and Curt Ellis, the filmmakers behind "King Corn." Web site: www.bigriverfilm.com

CHAPTER ELEVEN

epigraph: (Berkeley: University of California Press, 2005), p. 68.

And the Sokol-Blosser commitment: Susan Sokol Blosser, *At Home in the Vineyard: Cultivating a Winery, an Industry, and a Life* (Berkeley: University of California Press, 2006), p. 173.

so should we all also walk around with bared breasts: See Robert B. Edgarton, *Sick Societies: Challenging the Myth of Primitive Harmony* (New York: The Free Press, 1992).

"Results showed a trend toward": René Morlat, "Long-Term Additions of Organic Amendments in a Loire Valley Vineyard on a Calcareous Sandy Soil. II. Effects on Root System, Growth, Grape Yield, and Foliar Nutrient Status of a Cabernet Franc Vine," *American Journal of Enology and Viticulture* 59:4 (December 1, 2008), pp. 375-86.

"free-range landscaping": Lisa Napoli, "Weed-Whacking Goats Will Work For Food," *Weekend Edition Saturday* (July 10, 2010). Gwendolyn Bounds, "Free-Range Landscaping: Rent.a.Goat.com and Others Bring in Goats to Trim the Yard, Get Rid of Weeds," *The Wall Street Journal* (August 4, 2010), p. D1.

And states like Ohio and California: Erik Eckholm, "Farmers Lean to Truce on Animals' Close Quarters," *The New York Times* (August 12, 2010), p. A18.

"closed-loop system": See http://www.threemilecanyonfarms.com/our_best_practices/sustainable_dairy_practices.html

"Some of the world's most renowned wineries": From Stuart Smith's blog "Biodynamics is a Hoax" (http://biodynamicshoax.wordpress.com/, June 1, 2010)

CHAPTER TWELVE

epigraph: *The Works of Epictetus*, trans. Thomas Wentworth Higginson (Boston: Little, Brown, and Company, 1912): *The Discourses*, Book 1, Chapter 15: "What Philosophy Promises," pp. 55-56.

Index

Index